Living Off-Grid

Solar and Wind Power Generation

JD Weiss & Twila Weiss

ISBN: 978-1-964391-06-9

REBEL RANCH PUBLISHING

Please visit our website

rebelranchpublishing.com

Table of Contents

Dedication

To our great friends Sean and Seana, whose constant support and invaluable experience of generating off-grid electricity have been indispensable. Your entirely off-grid home has inspired and enlightened us in ways we never imagined. Thank you so much for sharing your expertise and passion with us.

Introduction

Let's face it, most of us take for granted the amazing things that make modern life so easy. We barely even notice the fridge humming away, keeping our food fresh, or the flick of a switch that chases away the darkness. And don't even get me started on the internet—it's like a constant lifeline to the world! But lately, more and more people are ditching the convenience of the power company for something different. They're trading the power company to generate electricity using solar and wind power. Why? It's because they're choosing a simpler, self-sufficient lifestyle—one that's all about living off the grid and saving money on electricity.

What Is Living Off-Grid?

Basically, it is all about being less reliant on the usual power lines and city services. That's the core idea of living off-grid! It's all about becoming more self-sufficient by generating your own electricity, often through clean sources like sunshine, wind, or even moving water. But living off-grid can go beyond just electricity. It can involve collecting your own water, managing waste in eco-friendly ways, and even growing your own food—which are subjects for our future books.

People choose this lifestyle for many reasons. Some folks are passionate about protecting the environment and want to reduce their carbon footprint. Others crave a simpler life: one that's less dependent on external systems and more focused on being resourceful and self-reliant. There are also those who are drawn to the freedom and independence that comes with making their own necessities.

Importance of Renewable Energy

People who need reliable and long-lasting energy sources are more likely to choose to live off the grid. Solar panels, wind turbines, and micro-hydro systems are getting better and less expensive, which makes it easier for people to make their own electricity. These are some of the main benefits.

Effects on the Environment

Your carbon footprint is smaller if you get your power from green sources like solar and wind power. Greenhouse gas pollution is cut down a lot when you don't use fossil fuels. This helps fight climate change and makes the world healthy. Living off the grid lets you control how you affect the world, which helps make the future more sustainable.

Having Your Own Energy

Energy freedom is one of the best things about living off the grid. If you make your own energy, you won't have to depend on the grid. You can feel safe and in charge of your energy needs. You don't have to worry about being without power—such as during a power outage—or looking for other energy sources.

Saving Money

Setting up a green energy system costs money at first, but it saves you a lot of money in the long run. You'll notice a big drop in your energy bills once your system is up and running. Off-grid life can save you a lot of money over time, so it's a good financial choice for the future.

Who Is This Book For?

Calling all curious minds and adventure seekers! Whether you're just daydreaming about living off-grid or ready to ditch the power lines for good, this book is your ultimate guide. It doesn't matter if you're a total beginner or have some experience under your belt—we've got you covered! Here's a sneak peek at what you'll find inside:

- **Planning makes perfect:** Before you unplug, let's get you prepped! We'll cover everything you need to know, from choosing the ideal location to maintaining the systems like a pro.

- **Power up with nature:** Ditch the grid and harness the sun and wind! We'll break down renewable energy systems, like solar panels and wind turbines, so you can become your own mini power plant.

Scope and Purpose of the Book

This book is your one-stop shop, from dipping your toes in self-sufficiency to a full-on lifestyle change! We'll guide you through every step to power your home with clean energy. Whether you're just looking to reduce your reliance on the grid or go completely unplugged, this book is packed with valuable tips and practical advice to make your off-grid adventure smooth sailing!

It's not all about wires and watts! We'll also explore the way of thinking that goes with living off-grid. You'll learn how to slow down, simplify your life, and reconnect with nature in a deeper way. Living off-grid isn't just about ditching the power company; it's about creating a life that feels more meaningful and intentional.

Think of this book as your personal off-grid GPS. We'll show you how to take charge of your life and environment—whether you're worried about the planet, crave independence, or just want things a bit simpler. Living off-grid can lead to a life that's both liberating and more sustainable, and this book is packed with practical tips and advice to guide you on your journey every step of the way.

Ready to unplug and go green? Let's dive into this adventure together and explore all the cool possibilities of living off-grid. Whether you're just starting out or looking to upgrade your setup, this book has all the tips and inspiration you need to build a self-sufficient, eco-friendly home. Welcome to the off-grid life—where you're in control.

Chapter 1:

Introduction to Solar Power

Installing Solar Panels

Solar power is a hot topic, and for good reason! It's a clean, renewable way to generate electricity, and it's becoming more affordable and efficient all the time. This chapter is your jumping-off point for going solar, especially if you're interested in living off the grid.

People have dreamed of harnessing the sun's energy for ages, but only recently has technology caught up. We'll take a quick look at how solar power has evolved—from early, not-so-great attempts to today's sunshine-powered workhorses: solar panels.

This chapter is here to give you a solid foundation in solar energy, whether you're curious about the history or ready to dive into the nitty-gritty of going off-grid. Get ready to explore the exciting world of solar power, where innovation meets environmental responsibility!

History and Development

Believe it or not, solar power has a surprisingly long history. Way back in the 8th century B.C.E., folks in Babylon were using sundials to tell time, based on the sun's position (The Editors of Encyclopædia Britannica, 2019). Fast-forward to Roman times—Emperor Flavius Domitian even taxed people for using sunlight in their greenhouses. (Can you imagine?) But the real turning point for solar power came much later.

In 1839, a French scientist named Edmond Becquerel discovered something amazing. He found that certain materials, like silver chloride, could create a tiny electric current when hit by light. This discovery, called the photovoltaic effect, became the foundation for solar power generation (Edmond Becquerel, 2019). Decades later, in 1873, an English engineer named Willoughby Smith was evaluating potential materials for undersea telegraph lines when he made the discovery that selenium could actually conduct light. The first solar cells were created from selenium in 1883 by American inventor Charles Fritts. Fritts had hoped that his solar cells could compete with Edison's coal-fired power plants, but they were only 1% good at turning sunlight into energy, making them not very useful. Still, it was a big leap forward!

The 20th century saw things really pick up for solar power. Albert Einstein's explanation of the photoelectric effect in 1905 helped scientists understand how light could be turned into electricity (Chodos, 2005). Then, in 1954, researchers at Bell Labs hit a home run; they created the first silicon solar cell that was powerful enough to be practical. This invention paved the way for the very first solar-powered satellite, Vanguard 1, which launched in 1958 (Perlin, n.d.).

The space race actually gave solar power another big boost. Satellites needed a reliable power source in space, and solar panels were the perfect fit. All the research and development to make solar panels better for space travel also made them less expensive and more efficient for here on Earth.

For a long time, though, solar panels were pretty expensive, so they weren't widely used. But as people grew worried about relying on fossil fuels and polluting the environment, interest in renewable energy sources—like solar power—skyrocketed. With government assistance and new technologies, the cost of solar panels came way down in the late 20th and early 21st centuries (HOW SOLAR PANEL EFFICIENCY, 2024). Today, solar power is a super affordable and fast-growing way to generate electricity!

Benefits of Solar Power

Solar power offers many benefits that make it an attractive energy source for off-grid living. Here are some key advantages.

Solar Is a Renewable Energy Source

Solar power can be used over and over again. Solar energy depends on the sun, a resource that will never run out—in contrast to fossil fuels, which are limited and get used up over time. Solar energy is a renewable and endless power source since it can be produced whenever the sun is shining.

Solar Energy Is Immensely Abundant

In just one hour, the sun gives Earth more energy than the whole world uses in a year. Because of this, solar power is a very stable and expandable way to get energy. No matter where you are, you can get enough sunlight to make a lot of solar power. This means that solar energy can be a reliable part of our energy mix.

Solar Technologies Are Getting More Efficient

Thanks to progress in technology, solar cells are now much more efficient. Photovoltaic (PV) cells today are better than ever at turning sunlight into power. It is still possible for solar panels to work better even when conditions aren't ideal because materials and manufacturing methods are getting better.

Solar Panels Are Becoming Less Expensive

In the last ten years, the price of solar cells has dropped by a huge amount. Improvements in manufacturing technologies have led to a decrease in costs. Because of this, solar power systems for homes and businesses are now more reasonably priced, which makes it easier for more people to use this type of clean energy. Solar energy is now one of the most affordable ways to produce power, thanks to falling installation costs and falling prices for solar technology.

Solar Life Cycle Generates Minimal Greenhouse Gas Emissions

Solar power systems generate electricity without emitting greenhouse gases during operation. The entire life cycle of solar panels, from manufacturing to disposal, results in far fewer emissions compared to fossil fuels. This makes solar energy a crucial part of efforts to reduce carbon footprints and combat climate change.

Not Dependent on Other Sources of Energy

A major perk of solar energy is that it doesn't rely on any other power grid. By harnessing the sun's rays directly, solar panels may produce power, eliminating the need for fossil fuels and making them less susceptible to changes in energy prices. In areas far from conventional power grids, where such lines do not exist, this independence is invaluable.

Electricity Bill Reduction

Power bills can be significantly reduced or eliminated with the installation of solar panels. You can lessen or do away with your dependence on grid power entirely if you generate your own electricity. Often, any extra energy that is made can be sold back to the grid, which lowers costs even more. The original investment in solar power systems can be recouped, over time, through savings on electricity bills, offering long-term financial benefits.

Basic Principles of Solar Energy

Solar energy captures the sun's power and turns it into electricity or heat for our homes and businesses. But how exactly does sunlight become something we can use? It all boils down to some clever science and technology. Let's explore the key ideas behind solar power to see how we can harness this amazing source of clean energy.

The Sun as an Energy Source

Our sun is like a giant furnace, constantly burning through a process called nuclear fusion. This releases a tremendous amount of energy that travels all the way to Earth as a wave of light, mostly the kind we can see and some we can't feel but keeps us warm (infrared). Even though Earth is far away, the sun showers us with enough energy to completely power our planet. And the best part? This sunshine is constantly refilled, making it a clean and endless source of fuel!

Photovoltaic Effect

Tiny particles of sunlight, called photons, hit a special material in a solar panel. These special materials, like silicon, act like little jump pads for electrons: tiny negatively charged particles. When a photon hits the material, it bumps the electron up to a higher energy level, kind of like kicking a soccer ball into the air. This creates a flow of electricity, just like moving water creates current in a river. This is the magic of the photovoltaic effect: the key ingredient that lets solar panels turn sunshine into usable electricity.

PV Cells

Solar panels are made up of many tiny building blocks called PV cells. These cells are like the power plants of the panel, and the main part is a material called silicon—kind of like the tiny gears and pistons that make a real power plant work. Each cell has a special two-layer design: a positive and negative side. This creates an invisible "force field" for electrons: tiny charged particles. When sunlight hits the cell, it acts like a bunch of tiny hammers, knocking electrons loose from their usual spots. With nowhere else to go, these free electrons zoom through the cell, creating a flow of electricity—just like water flowing through a pipe!

Types of PV Cells

Not all solar cells are created equal! There are three main types:

- **Mono:** Made from a single, perfect crystal of silicon, these are the most efficient at turning sunshine into electricity, but they can also be the priciest.

- **Poly:** These cells are made from lots of smaller silicon crystals, making them more affordable but a tad less efficient than their monocrystalline cousins.

- **Thin film:** These lightweight cells use a super thin layer of special materials. They're bendy and flexible, making them great for curved surfaces, but they don't quite pack the same punch as the crystal crew.

Solar Panel Construction

A solar panel has a team of tiny power plants working together. Each power plant is a PV cell, and to get the most electricity out of them, we need to connect them strategically. We do this by wiring them up in a special way—some in a row (series) and some in parallel. This lets us boost the overall power output—kind of like joining hands together to lift a heavy box.

But these little power plants need some protection from the elements! That's where the tough outer shell comes in. It's made of layers of special glass and plastic, like a superhero's force field. This keeps out moisture, dust, and anything else that might try to mess with the delicate PV cells inside. This shell also helps ensure the sunlight reaches the cells efficiently.

In the end, the whole team of mini power plants needs a strong foundation. That's where the frame comes in, usually made of sturdy aluminum. This frame acts like a skeleton, holding everything together and making it easy to mount the panel securely.

Inverters and Energy Conversion

Solar panels are great at turning sunshine into electricity, but there's a catch! The electricity they produce is called direct current (DC), which is like water flowing steadily in one direction. The problem is, most of our appliances and the power grid run on alternating current (AC)—Imagine water flowing back and forth. To bridge this gap, we use a clever device called an inverter. Think of it as an adapter that translates the DC language of the solar panels into AC language that our homes and businesses can understand. This way, the clean, solar-generated electricity can flow smoothly and power everything from your toaster to the lights.

Types of Inverters

Not all solar panels are created equal, and the same goes for the inverters that turn their DC output into usable AC power! Here's a breakdown of the three main types.

String Inverters

The workhorses of the solar world, these guys take the DC electricity from a whole string (group) of solar panels and convert it all at once into AC. They're a good choice for simple setups, but if one panel gets shaded, it can drag down the whole group's performance.

Microinverters

Talk about going solo! These mini-inverters are attached directly to each solar panel. They convert DC to AC right there on the spot; so, even if one panel is shaded, the others can keep pumping out power at full capacity. They're a great option for roofs with uneven shading or oddly shaped layouts.

Power Optimizers

Think of these as a happy medium between string inverters and microinverters. They work with a central string inverter, but each panel also has an optimizer attached. The optimizer fine-tunes the DC output from each panel before it reaches the inverter, squeezing out every last drop of power, even in less than ideal conditions.

Solar Thermal Energy

Sunshine isn't just good for making electricity! Solar thermal systems are another way to harness the sun's power but, instead of electricity, they capture heat. These systems use special panels, called solar collectors, to soak up the sun's rays. There are different types of collectors—kind of like having different tools for the job.

Flat-plate collectors are the most common—perfect for heating water in your home, especially in warmer climates. Evacuated tube collectors are like high-tech thermoses, trapping more heat and working better in colder areas. And finally, concentrating collectors use mirrors or lenses to focus the sun's power like a magnifying glass, generating super high temperatures for industrial uses and even large-scale power plants.

So, what can you do with all this solar heat? Lots of things! Homes can use solar thermal systems for hot water and even space heating, helping you ditch traditional energy sources and save money. On a bigger scale, solar thermal power plants use that intense heat to create steam, which then spins turbines and generates electricity.

Solar Energy Storage

One of the main things to consider with solar power is that the sun isn't always out and shining. This can be a challenge because we still need electricity at night or on cloudy days. But there are clever ways to store that sunshine power for later use.

Battery Backup

Solar systems can use batteries to store extra energy produced during sunny days. This way, you have a reserve to tap into when the sun goes down or the clouds roll in. There are different battery types, each with its own pros and cons. Some are more affordable, while others last longer or hold more power.

Heat Banks

Solar thermal systems can also store sunshine in the form of heat! They use special materials—like water, molten salts, or special waxes—that can change form to trap heat. This stored heat can then be used to generate electricity or keep you warm, even when the sun isn't shining.

System Integration and Optimization

Well, you're ready to harness the power of the sun. But how does your solar system connect with the bigger picture—the entire energy grid? It's all about careful planning and making sure everything works together seamlessly.

Sizing Up Your Sunshine Needs

The first step is figuring out how much solar power you'll need. This involves considering how much energy you typically use, how many hours of sunlight your area gets, and choosing the right number and type of solar panels. It's like picking the perfect outfit for your roof—not too big, not too small, just the right amount of sunshine-collecting.

Selling Your Sunshine Surplus

A cool perk of solar power is that you can become a mini energy producer. Many places have net metering programs. This means any extra electricity your solar system generates during the day can be sold back to the grid. It's a win-win; you get a credit on your electricity bill, and the grid gets a little cleaner, sun-powered boost.

Understanding these key concepts, from how solar panels work to how they interact with the grid, is important to getting the most out of this clean energy source. As technology keeps getting better, solar power is sure to become an even bigger player in powering our future.

Chapter 2:

Site Evaluation for Solar Power

Roof-Mounted Solar System

Before you jump into installing solar panels, it's important to find the perfect spot for them. This chapter will be your guide to figuring out if your location is a solar superstar. We'll explore the key ingredients for a happy solar system, from understanding how the sun moves in the sky to using cool tools to size up your roof.

By doing a proper site evaluation, you can ensure your solar panels are soaking up the most sunshine possible, which means they'll work better and generate more clean energy for you. So, get ready to find your sunshine sweet spot and make a smooth switch to solar power.

Assessing Solar Potential

To install solar panels, it's important to do some detective work and find the sunniest spot on your property. This is called a "solar site assessment," and it's like checking the weather forecast for your roof. In this step, different factors are looked at to see if a spot is good to generate solar energy.

Solar Radiation and Geographic Location

When it comes to solar power, one of the biggest factors is how much sunshine your panels can soak up. This depends heavily on your geographic location. Areas closer to the equator are bathed in more direct sunlight year-round, making them solar superstars. But don't fret if you're not exactly equatorial! Even places further north or south can be viable for solar energy, as long as they have two key things:

- **Good sun exposure:** Think of your roof as a prime tanning bed for your solar panels. The more direct sunlight they receive throughout the day, the more power they can generate.

- **Minimal obstructions:** Shade is the enemy of sunshine. Trees, buildings, or anything else casting a shadow on your panels can greatly reduce their efficiency.

So, how do we measure this sunshine strength? We use something called "solar radiation," which basically tells us how much sun energy hits a specific area. It's measured in kilowatt-hours per square meter per day (kWh/m^2/day). The higher the number, the more sunshine power your potential solar site has.

To get a good idea of a specific location's solar radiation levels, we can use two helpful tools:

1. **Solar radiation maps:** These maps show the average amount of sunshine different regions receive, giving you a general idea of your area's solar potential.

2. **Meteorological station data:** Weather stations track sunshine duration, which can provide valuable insights into the exact amount of sunlight your specific location gets.

By considering your geographic location and using these tools, you can ensure your solar panels are positioned in the sunniest spot possible, maximizing their power generation and making your switch to clean energy even more effective!

Site Orientation and Tilt

Once you've found the sunniest spot on your property, it's all about positioning your solar panels for maximum sun capture. Just like a sunflower turning its head towards the sun, solar panels work best when they're facing directly at the sun. In the Northern Hemisphere, this means facing true south. In the Southern Hemisphere, it's true north.

Tilting for the Seasons

The Earth's tilt throughout the year means the sun's position in the sky changes. To catch the most rays all year round, the angle of your solar panels needs to be adjusted based on your location's latitude. It's like tilting a picture frame to get the perfect lighting—but for sunshine instead of a camera flash!

Shading and Obstructions

Trees, buildings, or anything else casting a shadow can steal the sunshine your solar panels need to make power. Even a little shade on a small part of a panel can make a big difference.

To avoid this, we do a "shading analysis" before installing solar panels. This is like playing detective to find out where shadows fall throughout the day and year. We use special tools to track the sun's path across the sky, so we can place the panels in the sunniest spot possible. By outsmarting the shade bandits, your solar panels can work at their best!

Climate and Weather Patterns

The climate and weather trends are also very important when figuring out the solar potential. Solar energy can be produced more effectively in areas with lots of sunlight and clear skies. On the other hand, places that get a lot of clouds, heavy rain, or snow may have less solar energy.

Sunlight Hours

To figure out how much energy a solar power system could produce, you need to know how many hours of sunlight it gets on average each day. You can get this information from databases of solar radiation and past weather records.

Effects of Weather

Knowing the local weather trends can help you make a solar power system that can work in harsh conditions—like high winds, heavy snowfall, or extreme temperatures. This will make the system more reliable and last longer.

Ground and Roof Conditions

It's important to think about the site's physical state, whether it's on the ground or a roof. When the solar array is placed on the ground, the type of soil and how stable it is are crucial for the base. For installations on roofs, the roof's structural soundness and load-bearing capacity must be checked to make sure they can hold the weight of the solar panels and mounting hardware.

Studying the Soil

For systems that are mounted on the ground, a soil study can tell you if the location is good for supporting the solar panel array. If the soil is loose or unstable, you may need to use more engineering methods to make the foundation stable.

Roof Structural Integrity

For installs on roofs, a structural engineer should check the roof to make sure it can safely handle the extra weight of the solar panels, mounting equipment, and snow that might fall.

Legal and Regulatory Considerations

Before you install solar panels, it's important to understand the local laws and regulations. These can include things like zoning rules (restrictions on what you can build where) and building codes (safety standards for construction). You might also need permits or approvals from your local government to get the go-ahead for your solar project.

The good news is that many places offer incentives to encourage people to switch to solar power. These can include:

- **Permits made easy:** Some local governments might have streamlined permitting processes for solar installations, making it easier to get approval.

- **Solar savings:** Federal and state governments—along with some utility companies—offer rebates, tax credits, or other financial incentives to help offset the cost of installing solar panels.

Understanding Sun Path and Solar Angles

Remember how we talked about sunshine being the fuel for your solar panels? To get the most out of that fuel, we need to understand how the sun moves across the sky. This is called the sun's path. By knowing the sun's path and solar angles—the tilt of the sunlight hitting the panels—you can position your solar panels perfectly.

The Sun's Path

The Earth's a bit like a tilted merry-go-round as it travels around the sun. This tilt, along with the Earth's rotation, affects how the sun moves across the sky throughout the year. The sun's path is important for solar power because it determines how much sunlight hits your panels at different times.

Solar Declination

This fancy term basically means how high or low the sun is in the sky at different times of the year. It's like the sun's angle compared to the Earth's equator. The angle changes throughout the year, reaching its highest point in the summer and its lowest point in the winter, in each hemisphere.

Solar Noon

This is the moment when the sun is at its highest point in the sky for your location. In the Northern Hemisphere, this happens when the sun is directly south, and in the Southern Hemisphere, it's directly north. Knowing solar noon helps us position the solar panels to capture the most sunlight during this peak time.

Solar Angles

We've talked about the sun's path and how it affects solar power. But there's more to it! There are three key angles we consider when positioning the solar panels for maximum sunshine capture.

Solar Altitude Angle

Imagine a line drawn straight down from the sun. The angle between that line and the ground is the solar altitude angle. It's highest at solar noon—when the sun is directly overhead—and gets lower in the morning and evening. The higher the angle, the more intense the sunlight hitting your panels.

Solar Azimuth Angle

This is basically the sun's compass direction in the sky. We measure it in degrees clockwise, from true north. By understanding this angle, you can orient your panels to face the sun most directly throughout the day.

Solar Incidence Angle

This is the angle between the sunlight hitting your panels and a line perfectly perpendicular to the panel's surface. The closer this angle is to zero, the more directly the sun hits the panels, maximizing their efficiency and power output.

Seasonal Variations

The Earth is a bit like a tilted teacup as it spins around the sun. This tilt is why we have seasons, and it also affects how much sunlight your solar panels receive throughout the year.

Summer Sun

In the summer, the Earth's tilt means the sun follows a higher path across the sky. This results in longer days with more direct sunlight hitting your solar panels—perfect for generating power!

Winter Sun

During winter, the Earth's tilt makes the sun follow a lower path in the sky. Days are shorter, and the sunlight hitting your panels is less direct. This means slightly less solar power generation, but that doesn't mean solar isn't helpful in the winter!

Equal Day and Night (Equinoxes)

Twice a year, during the spring and fall equinoxes, the Earth's tilt is neutral. Day and night are roughly the same length, and the sun rises precisely in the east and sets smack-dab in the west. These equinoxes act like bookends, marking the changing seasons and the sun's path throughout the year.

Optimal Panel Placement

Think of your solar panels as catchers for the sun. If possible, place them so they face the sun to maximize its strength, especially at midday when it's at its highest point.

As a general rule, it's best to tilt the panels so they match the pitch of your roof. Typically, this angle is rather near to your geographic location on Earth (latitude). But if you can adjust the tilt, that's even better! You can adjust the angle of the panels so that they capture the lower sun rays in winter and the higher sun rays in summer. Some advanced systems can even move the panels around during the day so that they line up with the sun, like a sunflower.

Shadows are another thing to be wary of. Your panels' efficiency will decrease if they are shaded by trees or other structures. There are even regulations in place to prevent your neighbors from planting trees that would shade your home in certain regions! Other options are to make agreements, called easements, with your neighbors that they won't block your sun; in some places, building things that block solar panels might be against the zoning laws.

Tools and Techniques for Site Assessment

Before you install solar panels, be sure your home is a good fit for them. This entails taking a good look at your property to see how much sunlight it receives and whether anything is blocking it. This ensures that your solar panels work properly and that you save the maximum money on electricity.

Solar Pathfinders

Think of a tool that could tell you exactly where the sun will be, all day, every day of the year. That is a solar pathfinder! It lets you know if there are any buildings, trees, or other shadows that might keep your solar panels from getting enough sun. If you know this in advance, you may position your panels so that they receive the most amount of sunlight.

A solar pathfinder is basically a fancy sun predictor for your solar panels. It's like a crystal ball; rather than seeing the future, it shows you where the sun will be shining throughout the day and year.

The pathfinder features a glossy dome and a built-in compass. You position it where you want to install your solar panels and aim the compass north. The dome then mirrors everything around it—even the sun's passage across the sky. This allows you to determine if any trees, buildings, or other shadows will prevent your panels from receiving adequate sunlight.

Solar Assessment Software

The amount of sunlight that reaches your roof is an important consideration when installing solar panels since it determines how much electricity the panels can produce. This is where the solar assessment software comes in. If you own solar panels, these programs are like having an assistant.

Allow me to explain: Solar evaluation software calculates how much sunlight your roof receives using satellite imagery and weather data. Shadows cast by nearby trees or buildings, as well as the angle at which you tilt your panels, are all factors that are taken into account by means of complex mathematical algorithms.

With this data, the program can roughly estimate the amount of power your solar panels will produce. If you want to know if solar power is right for your house and how much money you can save on your power bill, this will tell you.

Popular solar assessment tools are SolarEdge Designer, HelioScope, and PVsyst. These applications can be quite beneficial in helping you create the optimal solar system for your requirements.

Geographic Information Systems (GIS)

Solar panels work best with lots of sunshine, but not every part of your property gets the same sun exposure. This is where GIS technology comes in. GIS is like a super map maker for solar panels.

Here's what it does: GIS takes fancy space pictures (satellite imagery) and other data to create detailed maps of your property. These maps show things like hills, trees, and anything else that might block the sun. This helps you see which areas get the most sunshine—the perfect spots for your solar panels.

GIS can also combine information from different sources, like weather data and maps of the land. This way, you get the most complete picture of how much sun your solar panels will actually get. By using GIS, you can plan the perfect layout for your solar panels to soak up the most sunshine and generate the most electricity.

Photovoltaic (PV) System Design Tools

PV system design tools help you plan how to set up and use your solar cells. They help you set up your panels in the best way possible based on your needs, the panels' efficiency, and the space you have.

Consider the local climate, the amount of available space on top of or below ground, and the amount of energy you'll need, before using these technologies. To get the most out of your solar panels, you can use these tools to determine the optimal orientation, tilt angle, and system size.

They can also help you put your panels in the best places to get the most energy and avoid shade. By experimenting with various setups in the tool, you can determine the most efficient design for your solar power system.

Drone Technology

Drones with cameras and sensors are being used more and more to check out sites. They can give you high-resolution aerial photos and thorough information about the site's terrain, plants, and possible shade.

Drones are a quick and easy way to see a lot of the area and get information and pictures that would be hard to get from the ground. They can also get to places that are hard to get to, giving them a full picture of the spot. Drones can also be used to make 3D models of the site, find problems with shading, and check the state of roofs and other buildings.

Handheld Solar Meters

Handheld solar meters, which are also called pyranometers or solar irradiance meters, find out how much sunlight is hitting a certain spot. These devices give real-time information on the intensity of the sun, which helps figure out how much solar power the location can produce.

The unit of measurement used by solar meters is watts per square meter, or W/m^2. You can figure out the site's average amount of solar radiation by taking readings at different times of the day and year.

Set up a solar meter to take readings at regular intervals at the spot where you want to put the solar panels. This information can be used to guess how much energy the solar system will produce and make sure the spot gets enough sun.

Site Surveys and Inspections

To find problems that might affect the solar installation, it is important to do a full site survey and inspection. This includes checking the actual state of the site—such as the roof's structure, the stability of the ground, and any electrical systems that are already there.

When installing solar panels on a roof, you should check the roof's state and make sure it can hold the weight of the panels and the hardware for mounting them. If fixes or reinforcements are needed, they should be done before installation.

When installing solar panels on the ground, the type of soil and how stable it is should be checked by a site surveyor to make sure the panels will be securely fixed. This could include testing the soil and doing technical reviews to find and fix any problems that might come up.

Regulatory and Permitting Considerations

Before you put up your solar panels, there might be a few hoops to jump through with the local authorities. This is to make sure everything is safe and up to code.

The key here is to understand the rules in your area. These might be zoning laws, which are basically rules about what you can build where on your property. There might also be building codes, which are like safety instructions for construction projects.

The good news is that some places even offer incentives or rebates for going solar! These are like discounts or rewards for choosing a clean energy source. So be sure to research the regulations in your area to see if there are any benefits to going solar.

Chapter 3:

Solar Power Equipment

Roof-Mounted Solar System Installation

Sunshine powers your home! But how does it all work? Get ready to dig into this chapter as we explore all the fascinating gadgets and technologies that can transform sunshine into electricity that you can use around the house. All these tools are part of the sunny squad, and they all do different things. To help you choose the best solar energy system for your house, this chapter will give you an overview of the most recent innovations in solar technology. We'll also talk about how these tools can be used together to give you sunshine power!

Types of Solar Panels and Their Efficiency

Solar panels are like the stars of the solar power show! They're the most noticeable part and the ones that do the main job: turning sunlight into electricity. But did you know there are different types of solar panels, and some work better than others?

This section will be your guide to the different solar panel choices. We'll break down the types and how efficient they are at capturing sunshine. By understanding this, you can pick the perfect solar panels for your needs and get the most power out of your sunshine team!

Monocrystalline Solar Panels

Solar panels that are monocrystalline are made from a single, long crystal structure, which is usually silicon. It is known that they work well and last a long time. In this process, a single crystal of silicon is grown and then cut into thin plates. This makes it easier for electrons to move, which leads to higher efficiency.

The efficiency of these panels is usually between 15% and 20%, but some more advanced types can go as high as 24% (Geotherm_Admin, 2017; Solar Optimum, 2023). In other words, they can make more power with the same amount of sunlight than other types of panels. They also work well when there isn't much light, like when it's dark or when they're in the shade. They also come in a sleek black color that looks attractive and is often chosen for home setups.

On the other hand, monocrystalline panels are more expensive up front due to the complexity and cost of their production. Over time, though, they may be less expensive because they work better and last longer.

Polycrystalline Solar Panels

Polycrystalline solar panels, sometimes called multi-crystalline panels, are developed by melting silicon crystals in a specific way. There isn't just one crystal in these panels, unlike monocrystalline ones. Instead, silicon pieces are melted and put into a mold. The mold is then left to cool and harden, making a solid block. To make the solar cells, this block is cut into plates.

The efficiency range of polycrystalline panels is slightly lower than that of monocrystalline panels, often falling between 13% and 16% (Geotherm_Admin, 2017). This is because the edges of the silicon crystals can make it take longer for electrons to move. But they are less expensive because the process of making them is easier and costs less.

Even though they are less efficient, these blue panels can be a good choice if space isn't an issue. To make up for their lower efficiency, they're best for big installations where more panels can be added.

Thin-Film Solar Panels

The process of creating thin-film solar panels involves covering a substrate—be it metal, glass, or plastic—with photovoltaic material in layers. Different types of materials are used to make thin-film panels, but cadmium telluride (CdTe), amorphous silicon (a-Si), and copper indium gallium selenide (CIGS) are some of the most popular ones.

The efficiency of thin-film panels is usually 10-12% lower than that of crystalline silicon panels (*Thin Film vs. Crystalline Solar Panels*, 2023). On the other hand, they have some benefits that make them good for certain uses. One of the best things about them is that they are flexible and light, so they can be used in a lot of different places, like on curved surfaces and in portable solar devices. In addition, they work well in places with high temperatures and little light.

Compared to crystalline panels, thin-film panels use less energy during production, which means they are more cost-effective. But because of the reduced efficiency, more panels are required to generate the same amount of power, which can lead to higher installation costs and more area being needed for the project.

Bifacial Solar Panels

An innovative kind of solar panel, bifacial solar panels can increase their total energy production by capturing sunlight from both sides. Usually, the backs of these panels are clear or made of two pieces of glass, which lets sunshine pass through and bounce back onto the back of the panel.

If you place bifacial panels correctly and the surface around them is reflective, they can produce 10% to 20% more energy than monofacial panels (Bifacial Solar Panels, 2024). When the panels are mounted high or on surfaces that reflect light well, like white roofs or sandy ground, they work especially well.

When it comes to the front, bifacial panels are about as efficient as monocrystalline panels. However, the extra energy they collect on the back makes them more productive overall. Even though bifacial panels are more expensive, the higher amount of energy they produce can make them worth the extra money.

Factors Affecting Solar Panel Efficiency

Sunshine loves cool and clear skies. Even though solar panels are great at capturing sunlight and turning it into electricity, a few things can affect how well they work. Just like you don't want to run too much on a hot day, solar panels work a little less efficiently when it's superhot. Making sure there's air circulation around the panels helps them stay cool and work better.

Even a little shade from a tree branch can be a big bummer for a solar panel. Shade can block some of the sunlight, making it harder for the panel to generate electricity. There are special tools to help reduce the impact of shade, but it's always best to try and keep your panels in full sun.

Advances in Solar Panel Technology

Solar panel technology is always evolving to become more efficient and affordable. As materials get better, like with perovskite solar cells, production costs will go down and efficiency will go up even more. Perovskite cells could have rates of more than 25%, and because they are easier to make, they could be made for a lot less money (Magdalin et al., 2023).

Putting solar cells into building materials is another exciting new idea. This is called building-integrated photovoltaics (BIPV). This innovation lets solar panels be easily attached to roofs, walls, and windows—making buildings look better and serving a purpose at the same time.

Choosing the Right Solar Panels

There are a number of considerations to make when choosing solar panels for an installation—such as your energy needs, available space, and budget. For those seeking long-term performance and efficiency, particularly in restricted areas, monocrystalline panels are the way to go. While thin-film panels are more adaptable for one-of-a-kind uses, polycrystalline panels are more cost-effective for bigger installations.

Last but not least, you need to know about the different types of solar panels and how well they work in order to choose the right solar power system for you. Solar energy is now a more practical and attractive option for a wide range of applications.

Mounting Systems and Installation

The mounting system is like the skeleton for your solar panels. It holds everything up and keeps it safe. This section will show you the different types of skeletons available, so you can pick the perfect one for your roof. We'll also talk about how to install them to make sure your sunshine team gets the most sunlight and can handle anything the weather throws at it.

Roof-Mounted Systems

The most popular type of solar installation, especially for homes, is one that is mounted on the roof. These systems are put right on the roof of a building, making good use of the room that is there.

There are two main types of systems that are placed on the roof: *tilted mounts* and *flush mounts*. With flush mounts, the solar panels are lined up straight to the roof, giving the system a low-profile look. This kind of mount works best on roofs that are already pitched in a way that lets the sun shine through. On the other hand, tilted bases let you change the angle of the panels, which lets you position them so that they get the most sunlight. For flat roofs, tilted mounts work best because they can be set to the best position for efficiency.

Installing systems that are placed on the roof takes more than one step. First, the roof's structural soundness must be checked to make sure it can handle the extra weight of the solar panels and mounting hardware. Roofs that are older or less strong may need to be strengthened.

Next, the mounting brackets are connected to the roof in a safe way. These bolts need to be lined up and spaced out correctly so that the weight is spread out evenly and the panels have a stable base. The rails are connected to the brackets once they are in place. This makes the frame that will hold the solar panels.

The solar panels are then put on the rails. Usually, screws or other fastening systems are used to keep them in place. There are electrical links, and to make sure it is safe, the system is grounded. The last step is to fix any holes or cracks in the roof to stop leaks.

Ground-Mounted Systems

Ground-mounted systems are best for bigger installations like commercial buildings or solar farms because they don't take up much room on the ground. These systems have many benefits, such as making upkeep easier and letting you choose the best angle and orientation for the panels without being limited by the pitch of the roof.

Ground-based systems come in two main types: *standard fixed mounts* and *pole mounts.* Concrete or steel piers hold fixed mounts to the ground, making the base strong and long-lasting. The panels are put up at a fixed angle that is usually best for the latitude of the site.

With a pole mount, the solar panels are put on rods that are driven into the ground. This kind of system can be changed so that the panels are tilted and oriented in different ways, which lets it work best during different seasons. Pole mounts are especially helpful in places where the amount of sunshine changes throughout the year.

For ground-mounted systems, the first step in the installation process is to prepare the spot by clearing it and making sure it is level. Electrical lines may need to be buried, and then the foundations are put in place. After the bases are set, the mounting frames are put together and attached to the bases.

Next, the mounting structures are used to attach the solar panels, and then the electrical connections are created. If the system is installed on the ground, extra security measures, like fencing, may be necessary to prevent theft or vandalism.

Tracking Systems

Tracking systems are a more advanced type of mounting system that can make a solar power placement much more efficient by following the sun's path throughout the day. Trackers come in two main types: *single-axis* trackers and *dual-axis* trackers.

Trackers with only one axis move the solar cells along that axis, which is usually east to west, to follow the sun's path each day. Compared to fixed mounts, this lets the panels get more sunlight during the day. Dual-axis trackers move the panels in two directions, east to west and north to south, so they can follow the sun's path more exactly throughout the day and year. Dual-axis tracks are the most efficient, but they are harder to set up and cost more.

Installing tracking systems is similar to installing ground-mounted systems, but it requires extra parts, like motors and control systems, to make tracking move. To make sure the tracking system works right and makes the most energy, it needs to be properly calibrated and maintained.

Considerations for Installation

Before you put up your solar panels, you need a strong and safe way to hold them in place. That's where the mounting system comes in! It's like a secure shelf for your sunshine team. There are a few things to consider when choosing a mounting system:

- **What's your roof made of?** Different roofs need different types of mounting systems.
- **Does it get windy where you live?** Strong winds require a stronger mounting system.
- **What's the weather like?** The system needs to handle rain, snow, and anything else Mother Nature throws your way.

There are also some rules you need to follow, kind of like building instructions. These rules make sure everything is safe and up to code. It's important to get the proper permits and have a professional electrician do the wiring to ensure things are done correctly.

The mounting system is an important part of your solar power system. Picking the right one and installing it correctly will make sure your sunshine team is safe, secure, and can work at its best for a long time!

Chapter 4:

Designing a Solar Power System

Ground Mounted Solar System

Ready to harness the sun's power? In this chapter, you will learn all you need to know to create a solar power system that meets your specific energy needs. Get the most out of your system by following our detailed instructions on how to determine your energy use, choose the best components, and optimize your setup. We'll talk about important things like system size, battery storage, and safety. Understanding these factors will help you create a solar system that produces the most energy, is reliable, and costs the least. This makes sure that the switch to solar power goes smoothly and successfully!

Calculating Energy Needs

A key part of creating a solar power system that works is figuring out how much energy you need. Checking the numbers correctly makes sure the system isn't too small or too big, as either could cause it to work less efficiently or cost more. The steps in this process are figuring out how much energy you use every day, determining how much power you'll need at peak times, and thinking about how much energy you will need in the future.

Evaluating Daily Energy Consumption

The first step is to evaluate your daily energy consumption. First, make a list of all the electrical tools and devices you use, along with how much power they use (in watts) and how many hours a day they are used on average. If you use a 100-watt light bulb for 5 hours a day, it will use 500-watt-hours (Wh) every day. Do this for all your electronics to get the total amount of energy you use in watt-hours each day. To have a better grasp, think about how usage changes with the seasons. Heating and cooling requirements can have a big impact on energy consumption.

Calculating Peak Power Needs

Next, think about how much power your home needs at its peak. "Peak power" means the most power that is needed at any given time. This is important because you don't want your solar power system to run out of power during these times. List the gadgets that will likely be used at the same time and add up their power ratings. For example, if you use a microwave (1,000 watts), a fridge (800 watts), and a TV (150 watts) all at the same time, you will need 1,950 watts of power at its peak.

Accounting for System Inefficiencies

You must also take the system's inefficiencies into consideration. None of these energy storage and conversion devices—batteries, inverters, and solar panels—are perfect. (In my opinion, you should add about 20% or so to your estimated energy needs to account for this.) If your daily energy demand is 10,000 Wh, for instance, it's wise to aim for a system that can generate 12,000 Wh per day to guarantee reliable performance.

Determining the Solar Power System Size

Once you know how much energy you use every day and at peak times, you can figure out what size solar power system you need. Just take the average number of peak solar hours that your location receives per day and divide it by your daily energy use. Peak sunlight hours rely on where you live and the time of year, but in most places you can expect 4–6 hours of sunlight each day. To illustrate, suppose your average daily energy use is 12,000 Wh and your location typically experiences 5 peak hours of sunlight. In this case, you would require a system that can generate 2,400 Wh/h (12,000 Wh/ 5 hours).

Considering Future Energy Needs

Besides the energy that is needed now, think about how much energy might be needed in the future. This could include plans to buy additional devices, change the size of the family, or switch to electric cars. If you include these things in your figures, you can help make sure that your solar power system will work well for many years to come.

Planning for Energy Storage

Another important thing to think about is energy storage, especially for systems that are not connected to the grid. Batteries hold extra energy that is made when it is sunny so that it can be used later or at night. To find out how much power you need, multiply how much energy you use each day by the number of days you want to store it. For instance, if you need a backup for 2 days and use 12,000 Wh per day, your battery will need to be able to hold at least 24,000 Wh.

System Sizing and Configuration

The heart of a powerful solar system lies in its size and setup. This means figuring out the perfect balance between the size of your solar panels, the inverter that converts their power, and any battery storage you might need. This ensures your system reliably generates and utilizes clean energy.

Calculating Daily Energy Consumption

First, figure out how many kilowatt-hours (kWh) of energy you use every day. Count up all the watt-hours that your usual electronics and machines use. To find out what size solar array you need, divide this number by the usual number of peak sunlight hours per day where you live. Say you use 30 kWh of electricity every day and get 5 hours of the best sunlight. Then you need a 6 kW solar array (30 kWh / 5 hours).

Considering Panel Efficiency

Next, think about how well the solar panels work. Efficiency rates for panels range from 15% to 20%, but can be higher or lower depending on the panel (Abeni, 2024). If you choose panels that are 18% efficient, you will need fewer of them to make the same amount of energy as panels that are less efficient. Divide the size of the whole system by the power of each panel to get the number of panels you need. Let's say each panel is 300 watts. To make a 6 kW setup (6000 watts/300 watts per panel), you need 20 panels.

Choosing the Right Inverter

The panels send out direct current (DC), which is changed by inverters into alternating current (AC), which your home needs. The size of the inverter you buy should match the total power of your solar panels. Get an inverter that can handle at least 6 kW for a 6 kW setup. If you want to add to your system later, it's smart to get an inverter that's a little bigger.

Planning for Battery Storage

When you're not connected to the power grid or need backup power during a power loss, batteries are very important. The energy you make during the day should be stored in your battery so that you can use it at night or when there isn't much sunlight. To find this, multiply how much energy you use each day by the number of days you want battery backup. For example, you need a 60 kWh battery bank to have power for two days and use 30 kWh per day. You should also think about the batteries' depth of discharge (DoD). The batteries will last longer if you only use a small part of their full power. You'd need a 120-kWh battery bank for a 60-kWh need with a 50% DoD (Foxtron, 2021).

Configuring the System

The solar panels must be connected in either series or parallel with each other to achieve the desired voltage and current levels during system setup. Series connections make the system's voltage higher, while parallel connections make the current higher. Your generator and battery storage should be able to get the power they need from the setup. In this case, if your inverter needs 48V, connect enough panels in series to get that power. Setting up the battery bank so that it matches the system voltage is another way to make energy storage and use more efficient.

Solar Installation Guidelines

To get the most out of your solar system, you need to make sure the installation is done perfectly. If you follow the right steps and best practices, your system will work at its best, be safe, and last a long time. It also makes sure you follow all local rules, which keeps you on the right side of the law.

Step 1: Site Assessment

- Before any installations are made on top of the roof, make sure the roof is structurally sound.
- To make sure systems mounted on the ground are stable, you need to do a soil study.
- Verify that the roof is strong enough to hold the extra weight and that the soil is acceptable for the foundation.

Step 2: Panel Placement and Orientation

- In the Northern Hemisphere, position the panels so they face true south—and in the Southern Hemisphere, tilt them so they face true north.
- Adjust the angle of tilt with relation to the latitude of the location.
- To get the most out of the sun, stay out of places that are shaded.

Step 3: Mounting System Installation

- To avoid water damage, make sure to use the right flashing and sealants when installing roofing materials.
- Pick mounting hardware that can resist corrosion and the weather where you are.
- Establish a solid foundation of concrete or steel for systems that are positioned on the ground.

Step 4: Electrical Wiring

- Choose connectors and cables that are resistant to ultraviolet light and of good quality.
- For all connections, be sure to follow the manufacturer's instructions, as well as any local electrical codes.
- To prevent electrical problems and lightning strikes, make sure the system is grounded correctly.

Step 5: Inverter Installation

- Overheating can occur if inverters are not installed in places with adequate ventilation.
- Make sure they can be easily accessed for inspections and repairs.
- For best results, follow the manufacturer's instructions for installation.

Step 6: Battery Storage System Installation

- Keep the batteries in an area that is not too hot, too dry, or too crowded.
- The energy storage demands will determine the size of the battery bank.
- For safety reasons, make sure to utilize the right fuses and disconnect switches.

Step 7: Regular Maintenance and Monitoring

- Keep solar panels clean by wiping them down with a damp cloth on a regular basis.
- Check the electrical connections and mounting tools regularly.
- Track performance and find problems quickly with the help of monitoring systems.

By following these steps, you'll pave the way for a solar system that's safe, efficient, and legal—delivering clean, reliable energy for years to come. Remember, regular maintenance is key to keeping your solar power shining bright!

Safety Considerations

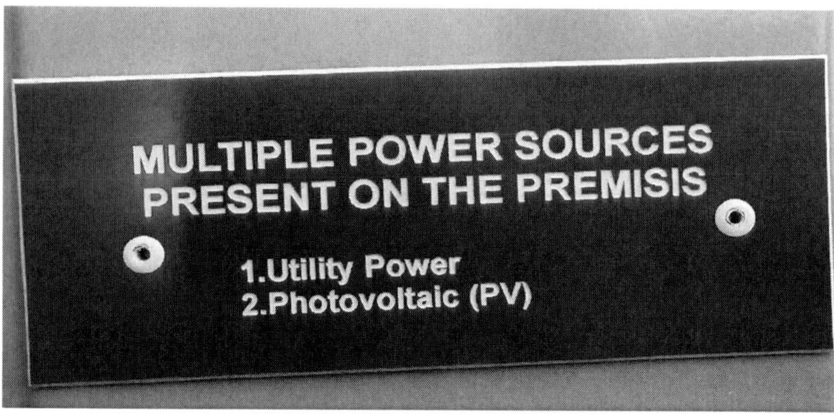

Power Warning Sign

When setting up and using a solar power system, safety is the most important thing. Taking the right safety precautions will protect the people working on the installation and make sure that the system is safe and reliable in the long run. A lot of crucial matters need to be thought about—like fire safety, electricity safety, and structural safety.

Because solar power devices use high voltages, electrical safety is very important. All electrical parts—like panels, generators, and batteries—must be handled and connected correctly during installation. To make sure that local electrical codes and standards are followed, electrical work should only be done by trained electricians. For electrical problems to stay away, you must use high-quality parts that are mentioned by UL and carefully follow the manufacturer's instructions.

The only way to keep the solar power system safe from electrical shocks and lightning is to ground it properly. Grounding means connecting the solar array's metal parts and holding structures to the ground. This gives electricity a safe way to flow in case of a fault. Using surge breakers can also protect the system from voltage spikes that come from lightning or other power problems.

Another important thing to think about is fire safety. Fires are much less likely to happen when electrical systems are designed and installed correctly. This means using wires, connectors, and safety devices like fuses and circuit breakers that are rated correctly. Overheating and fire dangers can be avoided by making sure all electrical connections are secure and corrosion-free.

This solar power system needs to be inspected and maintained on a regular basis so that any possible safety issues can be found and fixed. This means looking for signs of wear or corrosion, as well as connections that aren't tight and wires that are broken. For better effectiveness and less chance of electrical problems caused by dirt buildup, clean the solar panels every so often to get rid of dust and other debris.

Safety signs are necessary to warn and protect people who might come into touch with the solar power system. You can avoid accidents and be prepared for emergencies by clearly labeling areas with high voltage, disconnect switches, and emergency shutdown procedures.

Solar power system installers and upkeep workers must get training and education. With the right training, everyone will know what the risks are and how to best reduce them. This means knowing how to use personal protective equipment (PPE) correctly, how to handle electrical parts safely, and what to do in an emergency.

Chapter 5:

Maintaining Your Solar System

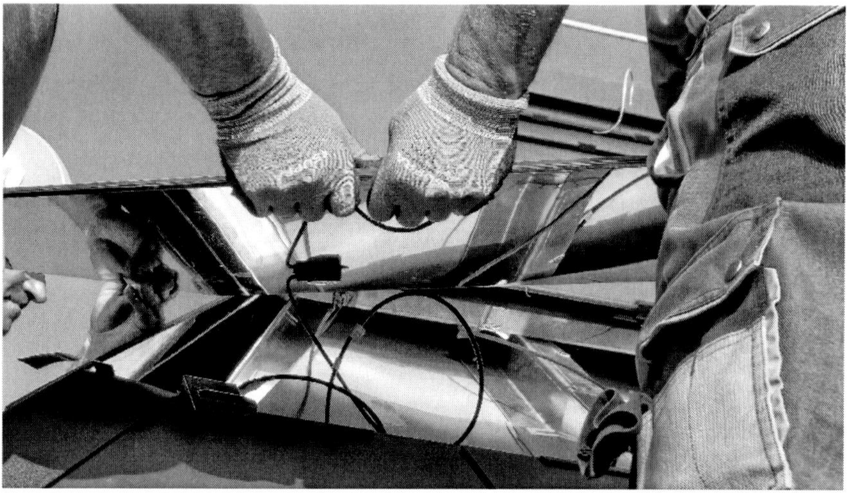

Solar Panel Installation

It's like sunshine in your pocket, all year round. This section equips you with the knowledge to perform regular maintenance, identify common glitches, and become a solar system troubleshooting whiz!

We'll cover the importance of routine checkups, cleaning strategies, and performance monitoring. Mastering these simple tasks can prevent headaches down the road, maximize your energy output, and extend the lifespan of your solar investment. So, dive in and learn how to keep your solar system shining bright—and ensure a continuous flow of clean energy for years to come!

Routine Maintenance Tasks

Don't let small gremlins turn into energy giants. Regular checkups and maintenance keep your solar system running smoothly and efficiently. By catching minor issues early on, you prevent them from snowballing into bigger problems and ensure your panels keep producing clean power for years to come.

Cleaning Solar Panels

Cleaning the solar panels is one of the most important regular maintenance chores. The panels can get dusty, dirty, or covered in bird droppings and other things that don't belong there. This makes them less effective because they block sunlight. Panels should be cleaned at least twice a year, or more often if you live in a place with a lot of dust, pollen, or smog. When you clean the panels, don't use rough materials that could scratch the surface. Instead, use a soft brush or sponge with water. In places with hard water, using boiled water can keep minerals from building up.

Inspecting for Physical Damage

You can train yourself to become a solar system detective by doing routine visual checks. Check the panels for any chips, cracks, or changes in color that could mean there is trouble below the surface. Don't forget to look for rust or loose nuts on the mounting hardware. For your solar powerhouse to stay structurally sound, you need to make sure that everything is tightly attached.

Monitoring System Performance

Keeping an eye on how well your solar power system is working can help you find any problems that aren't normal. A lot of new systems come with tracking software that shows how much energy is being made in real time. Check this data often to see if there are any big drops in energy output. These could be caused by shade, dirt buildup, or a broken part. If there are problems with your inverters or the grid, monitoring tools can also let you know.

Checking Electrical Connections

If you care about the reliability and effectiveness of your system, you must inspect the electrical connections. Electrical problems, overheating, or fire risks can be caused by connections that are too loose or rusted. Check all the wires and connections often, especially the ones at the solar panels, the inverter, and the battery storage. Make sure that every link is tight and that there is no corrosion on them. To keep good electrical contact and stop corrosion, put dielectric grease on all electrical connections.

Inverter Maintenance

The inverter is an important part of your solar power system because it changes the DC electricity from the panels into AC electricity that your home can use. Look for warnings or trouble messages on the inverter's screen on a regular basis. To avoid overheating, keep the inverter's cooling fins or fans clean and clear of obstructions and dust. Check the inverter's performance indicators on a regular basis to make sure it's running well. To keep the inverter in the best possible condition, follow the maintenance instructions given by the manufacturer.

Battery Maintenance

Don't let your battery bank become a power downer! For systems with battery backup, keeping those batteries in top shape is key. Give the terminals a regular check for corrosion; a quick cleaning can keep things running smoothly. Make sure the battery area is well-ventilated and free of dust bunnies and moisture; a clean environment is a happy environment for your batteries. Finally, use your system's monitoring tools to keep an eye on the battery's health and charge level.

Inspecting for Shading Issues

Watch out for sun stealers! Shadows can be sneaky energy vampires for your solar system. Make it a habit to check around your panels for any new shade culprits, like growing trees or nearby construction. Don't be afraid to trim back any vegetation that might be casting a shadow on your sunshine collectors. You can also use your system's monitoring data to become a solar detective. Look for patterns of reduced energy production at certain times of the day; that could be a clue that shade is hiding somewhere.

Reviewing Safety Systems

Make sure all the grounding connections are secure and shiny—no room for rust here! Take a look at your surge protectors too. If they look worn or have bravely defended against a power surge in the past, it's time to replace them for continued peace of mind.

Professional Inspections

DIY hero, but with a sidekick! While you can tackle many maintenance tasks yourself, don't be afraid to call in a solar specialist for a periodic checkup. A qualified technician can give your system a deep dive, examining everything from the electrical components to the structural integrity, making sure it's performing at its peak. These professionals can also spot hidden issues that might slip by during your regular cleaning and checks.

Common Solar System Issues and Solutions

Even the most reliable solar systems can hiccup sometimes. Don't worry, this section will equip you to troubleshoot common culprits that can steal your sunshine savings. By understanding these glitches and their fixes, you can keep your solar power system running smoothly and efficiently for years to come.

Decreased Energy Production

One problem that often happens with solar power systems is that they produce less energy than they used to. This could be due to a number of things—such as shade, dirt building up on the panels, or solar cells breaking down. Cleaning the panels often to get rid of dust, dirt, and other debris can make them produce a lot more energy. Also, make sure that nothing new is blocking the sun from hitting the panels, like trees that are growing or new buildings. Monitoring tools can help figure out where and when production drops so that action can be taken right away.

Panel Damage

Solar panels can get cracks or chips from things like heavy rain or snow, falling objects, or rough handling during installation. Damaged panels can make them less efficient or even cause them to fail completely. Look at the panels often for signs of damage. If damage is found, the panels must be replaced or fixed right away to keep the system working well as a whole. Using high-quality panels that will last and making sure they are installed correctly can lower the chance of damage.

Inverter Issues

Inverters are a vital part of the solar system—we all know it. Some common problems with inverters are that they get too hot, won't turn on, or don't work as efficiently as they should. To keep the inverter from getting too hot, make sure it is placed in a well-ventilated area and that its cooling fans or fins are clean and not blocked. Before you try to turn on the generator, make sure the connections and fuses are correct. Updating the inverter's firmware on a regular basis can also help fix problems with performance. If the problems don't go away, you might need to talk to the manufacturer or a trained technician.

Electrical Faults

Electrical problems, like connections that are too loose or rusted, can make systems work less well or even cause safety risks, like electrical fires. Check all the electrical connections often to make sure they are safe and free of corrosion. This includes the connections at the solar panels, the inverter, and the battery storage. Putting dielectric grease on joints can keep them from rusting and keep the electrical contact good. Also, to avoid electrical shocks and lightning strikes, make sure the system is grounded correctly.

Battery Problems

Problems that often happen with systems that store power in batteries are low battery capacity, burning, and not charging or discharging properly. Check the battery connections for corrosion often and clean them if they get dirty. Make sure that the battery case has good air flow around it and is dry and dust-free. Keep an eye on the battery life and charge level to spot problems before they escalate. If the problems don't go away, you might need to get new batteries.

Monitoring System Failures

A solar power system's performance and health can be greatly improved with the help of data collected by monitoring devices. If the tracking system fails, you might not be able to see how the system is working and problems might not be found right away. Make sure that all of the sensors and links work properly and that the firmware for the monitoring system is up to date. Reviewing the tracking data on a regular basis can help find patterns or outliers that point to deeper problems.

Weather-Related Issues

Weather extremes—like hail, heavy snowfall, or strong wind around it—can damage solar cells and other parts of the system. Putting up panels at the right tilt angle can keep snow from piling up and make wind resistance lower. Strong wind can damage systems if they are not properly secured and mounted with high-quality hardware. After bad weather, you need to check the system for damage and fix it right away if you see any.

Degradation Over Time

A solar panel's natural degradation over time causes it to become less efficient eventually. Every year, most panels lose between 0.5% and 1% of their power (Abeni, 2024). Degradation can be lessened by replacing old panels when they need to be. An extra layer of security can be added by picking high-quality panels with warranties that cover performance loss.

Upgrading and Expanding Your Solar Setup

Maybe your energy needs have grown, or you just want to squeeze every drop of sunshine out of your system. Upgrading can be a great way to boost efficiency, increase power production, and keep up with your growing energy demands. This process involves a few key steps: checking your system's current performance, picking compatible parts, and making sure everything integrates smoothly with your existing setup. We'll guide you through it all!

Evaluating Current System Performance

Take the time to evaluate how well your present solar power system is working before investing in any additions or modifications. You can learn a lot about your system's health from its performance data and monitoring tools. Consider things like energy production, system efficiency, and any problems that keep coming up. You can make better upgrade decisions when you've identified underperforming components or locations that need more efficiency in the system.

Determining Energy Needs

Before diving into upgrades, it's important to understand your energy appetite—both now and in the future. Think about any changes on the horizon—like new appliances, electric vehicles, or a growing family. By calculating your extra energy needs, you can size your upgrade perfectly to meet your demands. Remember, planning for future growth saves you time and money in the long run.

Selecting Additional Solar Panels

Adding more panels is a great way to boost your solar system! But to make sure everything works together smoothly, it's like finding a roommate that gets along with everyone. The new panels need to match the voltage and current of your existing ones; think of it as speaking the same language. On top of that, consider getting high-efficiency panels. They're like VIPs, squeezing the most energy out of limited space.

Upgrading Inverters

Your solar system's brain might need an upgrade too. As you add more solar panels, your inverter—the brains of the operation—might need a power boost to handle all that extra sunshine. Make sure the new inverter can handle the increased energy production and has all the bells and whistles. Modern inverters are like super smart assistants, constantly monitoring your system for peak performance and providing data that can help you with future maintenance.

Enhancing Battery Storage

For battery backup systems, expanding your battery bank is like having a bigger reserve tank for sunshine power. First, check the health and capacity of your current batteries. Do you need more juice to meet your energy needs? If so, consider factors like the new batteries' capacity, how many times they can be charged and discharged (cycle life), and, of course, compatibility with your existing setup. Adding new batteries might also mean upgrading your charge controllers—like getting a new faucet for a bigger tank. This ensures your batteries are charged and discharged efficiently, keeping your sunshine savings flowing smoothly.

Improving Mounting Systems

As you add more panels, make sure your mounting system can handle the extra weight. This might mean beefing up your existing structures or bringing in new mounting hardware—like building a bigger shelf for your growing trophy collection. Remember, proper placement is key for maximum power production. Think about shade—you don't want shadows giving your new panels a case of the blues! Adjusting their position a bit can help them soak up all the sunshine they deserve.

Integrating Smart Technologies

Level up your solar game with smart tech. Smart inverters, energy management systems, and fancy monitoring tools are like giving your solar system a brain and a sixth sense. These upgrades provide real-time data and insights, helping you use your sunshine power more wisely and catch any problems before they become headaches. Plus, they make it easy to add even cooler stuff in the future—like a supersized battery bank or a home automation system that works seamlessly with your solar. The result is a solar setup that's super efficient, responsive, and ready to meet all your energy needs!

Ensuring Regulatory Compliance

Upgrading your solar system is exciting, but don't forget to check in with your local authorities first. Getting the necessary permits and approvals is like getting a thumbs-up that your upgrade meets all the safety and performance standards. Following the rules keeps your system safe and legal, but there's more! Complying with regulations can also unlock cool incentives and rebates from local programs, if any are offered in your area. It's a win-win!

Professional Assistance

While you might be a solar whiz, for complex upgrades, consider calling in the professionals. These solar installation specialists can guide you through designing your system, picking the perfect parts, and making sure everything integrates smoothly. Plus, they'll ensure the installation is safe and efficient, saving you from any headaches down the road that might come from a DIY mishap.

Chapter 6:

Introduction to Wind Power

Old-Style Windmill

Are you excited to ride the wind to a greener future? Wind power, which uses the force of the wind to generate electricity, is an emerging renewable energy source. This chapter is your launchpad for understanding wind energy! We'll take off with the basic principles, exploring how wind turbines work their magic to convert wind into electricity. Then, we'll soar through the history of wind power, tracing its journey from ancient times to the cutting-edge turbines of today.

Buckle up, because we're also diving deep into the advantages of wind energy—like affordability, sustainability, and low environmental impact. By understanding how wind turbines work and the factors affecting their efficiency, you'll gain a strong foundation for exploring this exciting renewable energy source. Let's discover how wind power can transform the way we think about energy consumption!

History and Development of Wind Power

Wind power has a long history. It all started way back in ancient Persia (think modern-day Iran) around 500 to 900 C.E. (Howard, 2017). Back then, people used windmills with vertical axes to grind grain and pump water, especially for farms. Then, by the 12th century, Europe came up with horizontal axis windmills; these were way more efficient and became super popular, especially in countries like the Netherlands and England (The Editors of Encyclopædia Britannica, 2020). They used them for things like managing water and grinding grains.

For a while, windmills weren't as popular because of steam and fossil fuels, but they came back in a big way in the late 1800s, especially in the US. Farmers and ranchers used them for pumping water and even generating electricity in places where there weren't any power lines (US Department of Energy, 2019). This is when windmills started turning into the wind turbines we know today!

Wind power got a modern makeover in the 1900s! In 1941, the USA built a giant windmill (called a turbine) that could power a whole town for a short time (Shahan, 2014). This showed everyone that wind power could be big-time.

The 1970s oil crisis made people look for energy that wasn't dirty oil. Governments and companies put money into making wind turbines better and cheaper. Denmark became a wind power champion, building big wind farms and figuring out how to make them work well.

By the 1980s and 90s, wind turbines were getting much better at making electricity and cost less to build (Shahan, 2014). They even came up with new fancy blades and materials. This made wind power more affordable than ever before, so more and more places started using it.

Today, wind power is one of the fastest growing ways to get clean energy. New and improved turbines can power thousands of homes! We're even building wind farms out in the ocean where the wind is extra strong.

More and more countries are using wind power—especially China, the USA, and Germany. Wind power is becoming a big part of the world's energy supply, and it's helping us fight climate change. Scientists are still working on making wind power even better for the future!

Benefits of Wind Power

Wind power is a great way to make clean energy. It helps fight climate change by not using any fossil fuels that pollute the air. There are many reasons why wind power is so popular, and we can break them down into three main categories: good for the environment, good for the economy, and good for people.

Environmental Benefits

One of the best things about wind power is that it does not harm the earth as much as other types of energy. Wind energy is a clean option to fossil fuels because it doesn't release any greenhouse gases when it's used. This is very important in the fight against climate change because limiting global warming requires cutting down on carbon pollution.

Also, compared to the amount of energy they produce, windmills don't take up much space. Even though the towers and the equipment that goes with them take up land, the area around the turbines can still be used for other things, like farming or grazing. This two-use approach to land makes it more useful while causing the least amount of damage to current uses.

Also, wind power doesn't use water, while thermal power plants do. Thermal power plants need a lot of water to cool down. In places where water shortages are a problem, this is a big plus that helps with the long-term management of water supplies.

Economic Benefits

There are strong business reasons to use wind power. Thanks to technological developments and economies of scale, wind power is lowering its price tag to compete with more conventional energy sources. Wind power has become one of the most affordable ways to make new energy because its levelized cost of electricity (LCOE) has gone down a lot in the last ten years (AleaSoft, 2023).

Commercial wind power projects are also good for the economy because it creates employment. A diverse group of experts—including engineers, builders, technicians, and ecologists—make up the wind energy industry's workforce. This job creation happens at every stage of a wind project's life, from planning and building to running and fixing problems.

On top of that, wind power projects often help the local economy. Landowners who rent out their land for wind turbines can make money from them, and local governments get tax money from them. These influxes of money can help pay for community services, changes to infrastructure, and other public goods.

Energy security can also be improved by wind energy. Diversifying a country's energy mix and relying less on foreign fossil fuels can help it become more energy-independent. This not only keeps energy prices stable but also protects countries from supply problems and political unrest.

Social Benefits

Pollution is cut down by wind power, which is good for general health. Contaminants released by conventional fossil fuel power stations include particulate matter, sulfur dioxide, and nitrogen oxides; they can aggravate respiratory and cardiovascular conditions. In contrast, wind turbines make energy without releasing these harmful pollutants. This means that the air is cleaner and people's health is better.

The use of wind power also encourages new technologies and skill development. There is more research and development going on because the industry is growing, which means that turbine technology and efficiency are always getting better. This new idea moves the industry forward, making a positive feedback loop of progress and acceptance.

Communities are often very involved with, and interested in, wind power projects. A lot of the time, developers work with local groups to solve concerns and make sure that projects are good for the community. If people feel like they own, and are proud of, their town—they may be more open to renewable energy projects.

Long-Term Sustainability

Wind power is naturally environmentally friendly. While fossil fuels are limited and run out, wind is a resource that will never run out. Wind can be used to make electricity as long as it blows, making it a stable and long-lasting source of energy for the future.

Global environmental goals, like those set out by the United Nations' Sustainable Development Goals (SDGs), are also in line with wind power. Wind power helps with many SDGs—such as affordable and clean energy, fair work and economic growth, and climate action—by producing clean energy, lowering negative effects on the environment, and boosting economic growth.

Technological Advancements

Wind turbine technology keeps getting better! New windmills are more reliable and can make more power even when the wind isn't that strong. We're also seeing a rise in offshore wind farms, which are giant windmills built out in the ocean. This is a great spot for windmills because the wind is always stronger and steadier out there. Plus, offshore farms don't take up any space on land and can put their power right where it's needed near the coast.

Grid Integration and Storage Solutions

Wind power is a great addition to our energy mix, but it can be tricky to fit it all in. The good news is, scientists are coming up with smarter ways to manage the grid and store energy. For example, we can use giant batteries to save extra electricity generated on windy days, then use that power when the wind isn't blowing as strong. This way, we always have a steady flow of clean energy.

Basic Principles of Wind Energy

To create power, wind turbines convert the kinetic energy of blowing wind into electrical energy. This method turns wind into usable electricity by using basic engineering and physics concepts. It's important to understand these concepts if you want to know how windmills work and how they can be used to make renewable energy.

The Physics of Wind

The sun plays a big role in creating wind. The sun heats up different parts of the Earth unevenly, which makes some areas have more air pressure than others. This pushes the air from the high-pressure zones to the low-pressure zones, creating wind.

The strength of the wind is like its *punch*: The faster the wind blows, the more power it has to spin wind turbine blades. That's because wind power increases super quickly as the wind gets stronger, not just a little bit! This is why even a small increase in wind speed can make a big difference in the amount of electricity a windmill can generate.

Wind Turbine Components

Let's break down a windmill! There are a few important parts that work together to turn wind into electricity:

1. **Rotor blades**: These are the parts of the rotor that you can see the best. Most modern wind turbines have three blades, which are meant to catch as much wind energy as possible. Blades are shaped and spaced in a way that makes the most of aerodynamics.

2. **Hub**: The hub links the nacelle to the main shaft and the spinning blades. It sends the mechanical energy that the blades have collected to the engine.

3. **Nacelle**: This is the housing at the very top of the tower. It holds important parts like the engine, gearbox, and control systems. The nacelle is made to keep these parts safe from the weather and make upkeep easier.

4. **Gearbox**: Most wind turbines have a gearbox that speeds up the rotor from a slow speed to the faster speed needed by the engine. Some new engines, on the other hand, don't need a gearbox because they are direct-drive.

5. **Generator**: The generator takes the rotor's mechanical energy and turns it into electricity. Usually, it works by electromagnetic induction, which means that an electric current is caused by a changing magnetic field.

6. **Tower**: The hub and rotor blades are held up by the tower. In order to catch faster winds, which tend to be found higher up, it needs to be tall enough.

7. **Control systems**: These systems keep an eye on and control how the turbine works. They include systems for controlling blade pitch and yaw, and systems for stopping the turbine. They make sure the turbine works safely and efficiently by changing the blades' direction and pitch to get the best results.

Aerodynamics and Energy Conversion

How do windmills make electricity? It all starts with the blades. As wind blows over them, it creates lift, like an airplane wing. This lift force is what makes the blades spin. That spinning motion is carried down a shaft, which is like a long stick, to a part called a gearbox or directly to a generator. The gearbox, if there is one, helps the spinning speed up even more. Finally, the generator uses all that spinning to create electricity!

There's a scientific rule called the Betz Limit that says no windmill can capture all the wind's energy, no matter how good it is (*The Betz Limit*, n.d.). That limit is around 59% (Energy Education Team, n.d.). Luckily, modern windmills are pretty good at their job, capturing about 75% to 80% of that limit, which means they're very efficient at turning wind into electricity (Ranjbar et al., 2019)!

Wind Turbine Siting

Where do we put windmills to get the most power? The best spots are windy, open areas with steady breezes and not a lot of tall trees or buildings in the way. Places like coasts, flat plains, and hilltops are great choices because the wind blows there more consistently.

Before building a wind farm, scientists use special tools to measure how much and how often the wind blows in a certain area. This helps them figure out how much electricity the windmills can make and where to place them for the best results. It's like picking the perfect spot in your backyard to fly a kite—you want a wide open space with a strong, steady breeze!

Grid Integration and Energy Storage

Wind power is a great addition to our energy mix, but it can be tricky to fit it all in because the wind doesn't always blow! Unlike power plants that burn coal or gas, which can run constantly, wind turbines only work when it's windy. The good news is scientists are coming up with smarter ways to manage the grid and store energy.

One way is with something called a "smart grid." It's like a smart assistant for electricity, helping to balance how much power is coming in from wind turbines and how much people need at any given time. We can also use batteries to save extra electricity generated on windy days, then use that power when the wind isn't blowing as strong.

Environmental and Economic Considerations

Wind power is good for the environment in a few ways, like we discussed. First, it doesn't pollute the air like factories that burn coal or gas. This helps fight climate change. Second, wind farms don't need water to work, unlike some other power plants.

Wind power can help lower energy bills in the long run, and it creates jobs in all sorts of areas, from building commercial and home windmills to keeping them running. Of course, no energy source is perfect. Wind turbines can sometimes bother birds and bats, and they can make a whirring noise. But scientists are always looking for ways to make wind power even better for the environment and for people living near wind farms.

Chapter 7:

Site Evaluation for Wind Power

Picking the perfect spot for windmills is vital! This chapter will guide you through what to look for in a great location. We'll talk about things like the lay of the land (hills are good!), how the wind farm might affect the environment (we want to be good neighbors!), how strong and steady the wind blows (because strong, steady winds mean more power!), and how to measure all of this stuff.

We'll show you tools scientists use to find the windiest spots, like wind maps and special wind-measuring machines. By learning how to read this information, you can make the best decisions about where to put windmills on your property to get the most power. Choosing the right spot is like finding the perfect location for a kite-flying competition—open space, strong breezes—that's the recipe for success! This will set the stage for a home wind power project that works great!

Assessing Wind Potential

A key part of developing a wind power project is figuring out how much wind there is. This process looks at a lot of important factors—like wind speed, consistency, and the local terrain—to see if a spot is possible and how much energy it could produce. A correct evaluation makes sure that the chosen site will produce the most energy and lead to a successful project.

Wind Speed and Consistency

Wind speed is the most important thing to look at when figuring out how much wind there is. Due to the cubic link between wind speed and power output, wind energy production is very sensitive to wind speed. A small increase in wind speed can lead to a big increase in energy production. For this reason, it is essential to track wind speed precisely.

The main tools used to find out the speed and direction of the wind are anemometers and wind vanes. Usually, these devices are put on different heights of meteorological towers to collect data for at least a year. This information helps us understand the normal wind speed, how it changes, and how it fluctuates with the seasons. The World Meteorological Organization suggests taking measurements for at least a year to make sure the data is accurate and to account for changes that happen with the seasons (Reiss et al, 2021).

Consistency, or the steadiness of the wind, is just as important as wind speed. A place where the wind speed stays pretty steady is better than one where the wind gusts up and down all the time. Winds that don't change direction make it easier to plan how much energy will be made and keep rotor parts from wearing out.

Topography and Surface Roughness

Hills and ridges are your wind farm friends! The wind blows stronger the higher you go, away from the ground where things can slow it down. Imagine your kite flying high in the open field compared to getting stuck in the trees—that's the difference hills can make for wind power.

Besides hills, you also want to look for land that's flat and open, like plains or maybe even near a big lake. Flat land lets the wind whoosh through easily, without getting all choppy and slow. Forests and cities are like speed bumps for wind, slowing things down. So, the smoother and higher the ground, the stronger and steadier the breezes your windmills will get, which means more clean energy!

Wind Resource Mapping

Wind resource mapping is the process of making accurate maps that show the speed and direction of wind in a certain area. These maps are created by combining information from anemometers, weather stations, and sophisticated modeling tools. Wind maps show where the wind might blow and help you find places that might be worth looking into further.

Wind resource maps are very helpful in the early stages of site evaluation because they help developers narrow down the possible locations on your property, before more thorough measurements are made. These maps usually show average wind speeds, the direction of the wind that blows most of the time, and how it changes with the seasons. This gives you a full picture of the wind possibilities in the area.

Environmental and Regulatory Considerations

We've talked about finding a breezy spot with smooth, open land for your wind farm. But that's not all. We also need to consider three other things:

- **Keeping the wildlife happy:** Building a wind farm shouldn't disrupt the creatures who already live there. So, we need to do some checks to make sure the windmills won't bother any birds or bats. This involves something called an environmental impact assessment (EIA)—it's like a report card to see if the location is good for both wind power and the local animals. (This usually isn't necessary for home wind turbine installations.)

- **Following the rules:** Every area has its own laws about what you can build and where. So, before you put up any windmills, you'll need to check the zoning laws and land use regulations. You might also need permits from the local government. Think of it like getting permission to build a treehouse in your backyard—gotta make sure it's okay first! By following these steps, you can make sure your wind farm is a good fit for the environment and the law.

- **Economic viability:** Evaluating wind potential also involves considering its economic feasibility. This includes figuring out how much it will cost to develop the site, which includes installing the turbines and keeping them in good shape.

Long-term Data Analysis

Imagine you want to know how strong the wind blows at your beach house all year round. A quick check might not tell the whole story; maybe it's super windy in the summer but calm in the winter. That's why picking a windmill location requires looking at wind data for a long time, typically one to three years. This way, you get a better picture of how the wind blows throughout the seasons and year after year. The longer you look, the more accurate your prediction of how much electricity the windmills can create!

There are also cool tools that can help us measure wind data, not just those spinning anemometers you might see. Technologies like Light Detection and Ranging (LIDAR)—like radar, but with light—and Sonic Detection and Ranging (SODAR)—like radar, but with sound—can give us super detailed information about the wind, especially in tricky areas where traditional tools might not work as well. These fancy gadgets help us make sure we're putting windmills in the absolute windiest spots on your property for the most clean energy production.

Understanding Wind Patterns and Speed

Wind isn't always blowing in the same direction, and that's important for wind power. Just like you wouldn't fly a kite on a calm day, wind turbines need consistent breezes to spin their blades and create energy. Understanding how the wind behaves is key to placing windmills in the perfect spot.

Global Wind Patterns

The main thing that affects wind patterns, is how the sun heats different parts of the earth's surface. Differential heating makes pressure changes in the atmosphere, which lets air move from areas with high pressure to areas with low pressure.

Worldwide wind patterns—like the trade winds, westerlies, and polar easterlies—are created by the Coriolis effect, which is caused by the Earth's rotation. In the Northern Hemisphere, this deflects wind to the right—while in the Southern Hemisphere, it does the opposite.

As you move farther away from the poles, trade winds move from east to west, westerlies move from west to east, and polar easterlies move from east to west. These large-scale trends are necessary to understand how winds behave in different areas.

Local Wind Patterns

Mountains, valleys, and water all play a role in how the wind blows around your wind farm! These features can create their own little wind systems, making the wind stronger or weaker in different spots. Here's how:

- **Beach breezes:** The ocean and land heat up differently throughout the day. During the day, the land gets hot faster, and the warm air over it rises. Cooler air from the ocean then rushes in to take its place, creating a sea breeze. At night, the land cools down faster than the water, so the wind blows the other way: a land breeze!

- **Mountains:** Mountains are like giant wind ramps! During the day, the sun heats the mountain slopes, and the warm air rises up. Cooler air from the valley then flows in to replace it, creating a kind of mini wind tunnel. At night, the cool air settles back down into the valley.

Wind Speed Variation

Wind speed actually slows down the closer you get to the ground, kind of like how smoke gets blocked by a wall. But the higher you go, the less the ground can slow the wind down, so it gets stronger. This is why wind turbines have those tall towers; they reach up into the faster, stronger winds to make the most clean energy.

Measuring Wind Speed and Direction

The information gathered from anemometers and wind vanes is analyzed to find out the usual wind speeds, the direction of the wind that blows the most, and the patterns of change. his gives a solid foundation for evaluating wind resources.

Different Wind Speeds

Many times, a Weibull distribution is used to show how wind speeds are spread out in a certain area. This helps people figure out how likely it is that different wind speeds will happen. The Weibull distribution's shape and scale factors show how wind speeds are spread out around the mean speed. Wind power plants work best in places where the average wind speed is high and the distribution is small. This means that the winds are strong and consistent.

Predicting Wind Behavior

To predict how wind will behave, computer models and simulations are used to look at wind patterns and speeds by utilizing information from weather readings, topographical maps, and the state of the atmosphere. Predictive modeling helps find the best place and angle for turbines to produce the most energy.

Impact of Climate Change

Climate change can alter the patterns and speeds of the wind, which could make wind power projects that are already planned, or that are being planned, less likely to work. When world temperature patterns change, so can the systems that control atmospheric pressure, which can cause changes in how winds behave. Continuous tracking and flexible plans are needed to deal with these possible effects and make sure that wind power systems will last for a long time.

Tools and Techniques for Wind Site Assessment

Before building a wind farm, we need to do some homework to make sure it will work well. This process is called "wind site assessment." It's like checking the ingredients and oven temperature before you bake a cake—you want to make sure everything is just right.

Meteorological Towers

Meteorological towers, also called "met towers," hold up anemometers, wind vanes, and other weather tracking tools at different heights. These towers, which can be temporary or permanent, are put in places where wind farms might be possible. Met towers receive information about wind conditions at different elevations—such as speed, direction, temperature, and atmospheric pressure. This information gives a full picture of the wind conditions. Wind speed usually goes up with height because there is less surface drag at higher elevations.

Remote Sensing Technologies: LIDAR and SODAR

Regular wind measuring tools are great, but what if the windiest spot is on a super tall mountain? That's where LIDAR and SODAR come in. These are fancy tools that use science—LIDAR with lasers and SODAR with sound waves—to measure the wind from far away, even way up high where traditional wind measuring towers can't reach.

Think of LIDAR as a super powerful flashlight that can measure how fast the wind is blowing and which way it's going. SODAR, on the other hand, uses sound waves like a whistle to do the same thing. These tools are especially helpful in tricky areas with mountains or rough terrain, where putting up a regular tower might be difficult. By using LIDAR and SODAR, we can get a super detailed picture of how the wind blows across the entire area, helping us pick the absolute best spot for windmills to create the most clean energy.

Wind Resource Mapping

These maps use information from tall towers that measure wind, special tools that use lasers and sound waves (like fancy flashlights and whistles!), and even computers to show how fast and which way the wind blows in different areas. The windier the area, the brighter the color on the map.

These wind resource maps are like a goldmine for people looking to build wind farms. They can quickly see which areas have the strongest and steadiest winds, which is what windmills need to spin their blades and create clean energy. There are also cool computer programs called Geographic Information Systems (GIS) that can take all this wind data and put it on a map, along with other information, like mountains and bodies of water. This helps pick the absolute best spot for windmills, making the planning process smooth.

Computational Fluid Dynamics (CFD) Modeling

Remember how we talked about hills and mountains being good wind farm buddies? Well, to really understand how the wind will blow around those hills, you've got to use super powerful computers and a process called CFD modeling. Consider it a fancy wind tunnel simulation.

CFD modeling is like showing the wind a tiny model of the land and watching how it flows around hills, valleys, and any other bumps. By doing this, you can predict exactly how fast and which way the wind will blow at different spots. This helps you place the windmills in the perfect places to avoid wind shadows—areas where other hills block the wind—and turbulence—bumpy wind that can be hard on the windmills.

CFD modeling is especially helpful in areas with lots of hills and mountains because it can account for all those extra twists and turns the wind might take. By using this supercomputer simulation, you can make sure the wind farm gets the most wind possible to create the maximum clean energy. CFD modeling is primarily used for commercial wind farms, but is listed here for informational purposes.

Wind Data Analysis

Once you've collected a bunch of wind data from those tall towers, lasers, fancy sound machines, and other tools, it's time to turn it into something useful. Here's how you'll analyze all that information:

- **Average wind speed:** This is kind of like the wind's usual speed. The faster it blows on average, the more electricity your windmills can create!

- **Windy days vs. calm days:** Also look at how often the wind blows strongly, weakly, or just right. You want a good balance—enough strong winds to generate power, but not so much that it gets too crazy.

- **Bumpy rides:** Sometimes, the wind can be gusty or choppy, like a bumpy bike ride. We call this "turbulence intensity," and smoother wind is better for windmills to spin steadily.

- **Slow down at the bottom:** Remember how we talked about wind shear, where the wind gets stronger the higher you go? Consider this when placing windmills to make sure they get the most powerful breezes.

By looking at all these things, you can predict how much electricity your wind farm can create and avoid any areas with super strong winds or gusts that might be too much for the windmills to handle. It's like checking the weather forecast before a picnic—you want to know if it will be a gentle breeze or a total blow-out!

Chapter 8:

Wind Power Equipment

Horizontal Axis Turbine

Let's get down to the nuts and bolts of wind turbines. These are the amazing machines that turn wind energy into clean electricity. This chapter will take you inside a windmill and show you all the important parts. Understanding how all these parts work together is key to keeping wind turbines running smoothly and efficiently. We'll also explore some cool new wind turbine designs that are pushing the boundaries of wind energy. Let's get down to it then, shall we?

The Types of Wind Turbines and Their Efficiencies

The most important part of turning wind energy into power is the wind turbine. They are roughly divided into different types based on how they are oriented, how they are built, and how well they work. Understanding these differences is important for choosing the right turbine for a job and getting the most energy out of it.

Horizontal Axis Wind Turbines (HAWT)

Have you ever seen a giant windmill spinning in the distance? That's likely a Horizontal Axis Wind Turbine (HAWT), the most widely used type. These turbines look like giant propellers with blades that spin horizontally on a tall tower. Unlike some windmills, HAWTs don't need to be pointed into the wind; they can swivel their entire body, called yawing, to catch wind blowing from any direction.

Design and Operation

Earlier, we discussed the components of a turbine; here's how they work together. The first thing you see are the HAWT's giant spinning blades, called the rotor. These blades are connected to a central hub, which spins a shaft running through the tower. Inside the tower's head sits the nacelle, like a high-tech engine room. Here, a gearbox boosts the slow spin of the rotor into high-speed rotations. Finally, a generator takes this super-fast spin and converts it into the electricity we use! Many HAWTs even have smart controls that can adjust the angle of the blades for the best performance and safety, depending on the wind conditions.

Efficiency

Think about a machine that could turn up to half of the energy from the wind into power! That's the impressive feat of HAWTs. This is possible thanks to a mix of smarts and placements.

To begin, HAWTs can catch wind from any direction. They don't have to be pointed into the wind all the time like old windmills did. They can turn their whole bodies around to catch breezes from any direction.

The second thing is that HAWTs are very tall. It's not just for show! They can get stronger, more stable winds because they are higher than the bumps and waves on the ground. They can turn more wind energy into power with this steady flow.

All of these traits work together to give HAWTs a great conversion rate, usually between 35% and 45%, but sometimes as high as 50% ("Horizontal Axis Wind Turbines," 2020). In other words, they can get a lot of useful energy from the wind.

Applications

HAWTs are workhorses for both land and sea. On land, you'll find them in giant wind farms across vast, flat areas and in home installations. These open spaces offer smooth sailing for the wind, allowing HAWTs to capture its energy efficiently. The key here is consistency—strong, steady winds are what these turbines crave.

But HAWTs aren't afraid to get their feet wet! Offshore wind farms are becoming increasingly popular. Here, turbines stand tall in coastal waters, where the wind blows even stronger and more reliably. With fewer obstacles in the way and steadier gusts, offshore HAWTs can churn out even more clean energy.

Vertical Axis Wind Turbines (VAWT)

Have you ever seen a wind turbine that looks like a giant eggbeater spinning on its side? That's a Vertical Axis Wind Turbine, or VAWT for short. Unlike the horizontal windmills you might be used to, VAWTs don't need to chase the wind. Their secret lies in their design—the blades spin on a vertical axis, no matter which way the wind blows. This makes them a great choice for areas with unpredictable wind directions.

Design and Operation

VAWTs, unlike their horizontal cousins, come in a few different flavors. The most popular are the Darrieus and Savonius designs:

- **Darrieus:** Imagine an eggbeater on its side, spinning in the wind. That's the basic idea behind a Darrieus VAWT. These turbines have curved blades that connect to form a cylindrical shape.

- **Savonius:** This design is all about simplicity. Picture two or more scoops catching the wind like a child's pinwheel. Savonius VAWTs are easier to build and maintain, but not quite as efficient as Darrieus turbines.

No matter the design, all VAWTs share a similar core. The wind spins the blades, connected to a central shaft that runs down to the base of the turbine. There, a generator waiting at the bottom converts that spin into electricity.

Efficiency

VAWTs are champions of adaptability, but they don't quite match HAWTs in terms of pure power output. This difference comes down to the way the blades move. While HAWTs do better in isolation, they decrease in energy output when in close proximity to other turbines; VAWTs increase efficiency by working within close range of others.

Imagine pedaling a bike uphill. That's kind of what VAWT blades do, sometimes. They experience more drag and stress as they rotate, affecting efficiency. However, VAWTs have their own uniqueness. They don't need to care about wind direction. This makes them perfect for windy areas with unpredictable gusts, like bustling cities. So, while they might not be the efficiency kings, VAWTs shine in their versatility.

Applications

VAWTs are the adaptable wind warriors of the turbine world! Here's why they thrive in unexpected places:

- **City Slickers:** Unlike their direction-fussy horizontal cousins, VAWTs can handle the unpredictable gusts that whoosh through city streets. This makes them ideal for rooftop installations in urban areas.

- **Off the Grid:** Forget miles of power lines! VAWTs can function in remote locations with their ability to harvest energy from variable winds.

- **Small but Mighty:** Need a clean energy solution for a smaller scale project? VAWTs are a great choice. Their simpler design makes them perfect for powering homes or even small businesses.

What's the secret to their versatility? It all boils down to maintenance and wind patterns. With key components accessible at ground level, VAWTs are a breeze to keep running smoothly. Plus, their unique design allows them to capture energy even when the wind changes its mind every five seconds.

Direct-Drive Wind Turbines

Have you ever heard a grinding noise coming from a machine? That might be the gearbox, a part that can wear down over time. Direct-drive wind turbines ditch the gearbox altogether! Instead, they connect the rotor directly to the generator. This sleek design offers two key advantages:

1. With one less moving part, there's less that can go wrong. This reduces maintenance needs and keeps the turbine running smoothly.

2. By eliminating the gearbox, the most failure-prone component in a traditional turbine, direct-drive designs boast increased reliability. They're built to endure and keep churning out clean energy for longer.

Design and Operation

While traditional turbines rely on a gearbox to crank up the speed of the rotor, direct-drive turbines take a different approach. They have no gearbox. Instead, the generator itself is specially designed to handle the slower rotational speeds coming from the rotor. These special generators come in two flavors:

- **Permanent magnet:** These generators use powerful magnets to create the electricity. They're efficient and require minimal maintenance.

- **Electrically excited synchronous:** This type uses electricity to create a magnetic field. They offer more control over the power output but might require slightly more maintenance.

Efficiency

The conversion efficiencies of direct-drive turbines are about the same as, or slightly better than, those of geared HAWTs. They are also very reliable. Because they are less complicated mechanically, they last longer and work better over time, saving money on repairs.

Applications

Direct-drive windmills can be used in wind farms both on land and at sea. Because they are reliable and don't need as much maintenance, they are great for installations that are far from shore, where it can be hard to get to for upkeep.

Small Wind Turbines

Not all wind turbines are giants! Small wind turbines are designed for a more personal touch, providing clean energy for homes, farms, and even small businesses. These mini powerhouses typically generate a few hundred watts to several kilowatts, perfect for meeting the needs of smaller-scale applications.

Design and Operation

Both HAWTs and VAWTs are suitable for use as small wind turbines. They are usually put on shorter buildings, which makes them easier to keep up. The purpose of these turbines is to provide electricity for homes, farms, and small businesses. They can work on their own or with solar power and battery storage.

Efficiency

Overall, small wind turbines are not as efficient as bigger commercial turbines, but they can still provide a large amount of energy for the purposes they were designed for. To get the most out of them, they must be installed and positioned correctly.

Applications

Small wind machines are great for places that aren't connected to the power grid, like rural areas and places that don't have easy access to it.

They use renewable energy to lower the cost of power and make the country more energy-independent.

Innovations in Wind Turbine Technology

New developments in wind turbine technology keep making them more efficient, less expensive, and able to be used in more places. Improvements in turbine performance come from changes in blade design, materials, and aerodynamics. At the same time, improvements in stability and downtime are made by improvements in control systems and predictive maintenance.

New technologies, like floating wind blades and airborne wind energy systems, are making it possible to use wind energy in places that weren't possible before. Floating wind turbines, which are attached to the seabed but not fixed, can be put in deeper water where the winds are stronger, which makes it possible to produce even more energy.

Tower and Foundation Systems

Wind turbines might be the stars of the show, but they wouldn't be reaching for the sky without a strong supporting cast. The towers and foundations are the hidden giants of wind farms, playing a crucial role. The nacelle and rotor blades are held up by these structures, which raise them to the best heights where wind speeds are higher and more constant.

Things like the size of the turbines, the conditions of the spot, and the loads that come from the environment affect how the towers and foundations are designed and built. This part goes into detail about the different kinds of tower and base systems, as well as the design rules and building methods used for them. It shows how important they are to wind energy projects.

Lattice Towers

The skeleton of a lattice tower is made of steel or aluminum and looks like a truss structure. These towers are less expensive and lighter than tubular steel towers, so they can be used for smaller windmills in the back yard and places where transport is hard. But lattice towers need more upkeep because their joints and links are out in the open and can rust or wear down over time.

Design Principles

- **Structural Integrity**
  - Structural stability is the most important thing to think about when designing wind turbine towers. Towers have to be able to handle a lot of different loads—such as wind forces, rotor movements, and the weight of the nacelle and blades. To make sure the tower can withstand harsh weather and operating strains over time, engineers employ finite element analysis (FEA) to model these loads and optimize its design.
- **Aerodynamic Efficiency**
  - The shape of the tower also has an effect on how well the turbine moves through the air. Keeping the tower's aerodynamic drag as low as possible cuts down on energy loss and boosts total efficiency. To do this, forms that are smooth and streamlined, as well as materials with a smooth surface, are best.
- **Modularity and Transportability**
  - Modularity and mobility are critical things to think about. The towers are made in parts that are easy to move and put together on-site. It is possible to build taller towers by adding more modular parts, which makes the building process scalable.

- **Maintenance and Durability**
    - Windmills need to be maintained and last a long time in order to be useful in the future. The materials used are chosen to last longer by resisting corrosion, wear, and damage from the surroundings.

Foundation Systems

The foundation system holds the wind turbine firmly to the ground, keeping it stable and spreading the weight across the soil or rock below. What kind of foundation system is used relies on things like the soil, the size of the turbine, and the loads in the environment.

Spread footings, which are another name for shallow supports, are used where the ground is stable and strong. The big concrete base of these foundations spreads the weight over a large area, which lowers the ground pressure. Shallow foundations are easy to build and don't cost much, so they work well for smaller wind turbines and good soil conditions.

Construction Techniques

Building wind turbine towers and supports is a process that has several steps that must be done carefully to make sure the structure stays strong and people are safe:

- **Site preparation:** Clearing the area, leveling the ground, and testing the dirt to find the right foundation type are all parts of site preparation.

- **Foundation construction:** The way the foundation is built depends on the type that is picked. For shallow foundations, the digging part is followed by putting in the reinforcements and placing the concrete.

- **Tower assembly:** Putting together a tower starts with putting in the base part. Then, each section is lifted and bolted to the one below it.

Chapter 9:

Designing a Wind Power System

To make sure that a wind power system works well and is reliable, it needs to be designed with a comprehensive view. This chapter talks about some of the most important parts of designing a wind power system. For example, it talks about how to evaluate the site, pick the right turbine, and figure out the right system size. We will talk about the technical parts of matching wind resources with the right turbine types, getting the most energy out of the wind, and connecting the system to the power grid without any problems.

There will be clear directions about environmental and regulatory factors, which will ensure compliance and lessen any negative effects. You will fully understand how to create a dependable and useful wind power system that is tailored to a site's specific conditions and energy needs after reading this chapter.

Calculating Wind Energy Needs

The first and most important stage in developing a practical home wind power system is determining the amount of wind energy that will be required. Finding out how much energy is used up, the peak power needs, and the site's wind resources are all parts of this process. Homeowners can construct an efficient system that satisfies their energy needs by precisely evaluating these aspects.

Assessing Energy Consumption

The first thing that needs to be done when planning a home wind power system is figuring out how much energy the whole house uses. Start by making a list of all the electrical items you have, how much power they use, and how many hours a day they are used on average. Example: a 1,000-watt device that is used for 5 hours a day utilizes 5,000 watt-hours (Wh) every day. The total daily consumption can be found by adding up how much energy all of the gadgets use each day. To get a more true picture, look at how energy use changes with the seasons, especially if heating or cooling is a big deal. After this careful review, the planned system will be able to meet the family's energy needs all year long.

Peak Power Requirements

The next important step is to figure out how much power is needed at peak times. This is the most power that the system needs at any given time and is very important for making sure it can handle the most loads. For example, the highest power requirement is found by adding up the wattages of all the high-power devices that are being used at the same time. Making sure the wind machine can meet this demand keeps the power from running out when it's needed the most. This is a very important thing to think about for home systems, where many machines are often running at the same time.

Understanding Wind Resources

The amount of wind at the spot has a big effect on how much energy a wind turbine produces. Look at past wind data, which is usually gathered over a year, to find out the average wind speed and how it changes over time. Installing wind vanes and anemometers at different heights gives exact readings. One way to determine how often various wind speeds occur is by looking at the distribution of wind speeds, which is typically represented by a Weibull distribution. To choose the right turbine and estimate how well it will work, you need to understand these trends.

Battery Storage Consideration

When designing off-grid or battery-storage systems, it's crucial to figure out how much space is needed in the batteries to store extra energy made during times of high wind for use during times of low wind. The battery should hold enough power to meet the needs for the expected amount of time that is without enough wind. For example, if the battery bank needs to hold at least 90,000 Wh of power and daily use is

30,000 Wh, the battery bank should be able to last for three days. If you plan your battery storage well, you can keep getting power even when the wind isn't blowing.

Grid Integration

When figuring out how to connect a wind turbine to the grid, it is essential to compare the energy needs of the grid and the wind turbine. Depending on where you live, net metering rules can change how extra energy is charged, which can change how the whole system is designed. The most efficient and cost-effective way to get the most out of the wind power system is to make sure it is the right size to balance output and consumption. With the right grid integration, homes can get power from both their own solar and wind power generators and the power grid, which saves them money and makes the system work better.

System Sizing and Turbine Placement

When it comes to building a wind turbine system for your home, two things are key: getting the size right and putting it in the perfect spot. Imagine a car that's too small for your family. A wind turbine that's not powerful enough is much the same! By properly sizing your system, you can ensure it generates enough electricity to meet your home's energy needs. There will be no more relying on the grid or running out of power on a windy day.

Likewise, think of your turbine like a giant wind catcher. The better the placement, the more wind it can grab and convert into clean energy. Optimal placement also considers things like noise and shadow flicker on your property and that of your neighbors.

By getting the sizing and placement right, you'll have a wind power system that's efficient, keeps your home powered, and minimizes any impact on the surrounding environment. It's a win-win for you and the planet.

System Sizing

The first step in system size is to figure out how much energy the whole house uses. To do this, you need to figure out how much energy all of your electrical appliances use every day. For instance, if 30,000 watt-hours (Wh) are used every day, the wind power system must be able to produce that much.

As a precaution, I recommend adding a little to the amount of total energy needed to account for changes in wind speed and possible errors. This will help, when deciding what size turbine to purchase, to cover any losses that may occur, making sure that the system can safely meet demand even when the wind isn't blowing.

The next step is to do a study of the wind resources. This means looking at wind speed data that has been received from anemometers for at least a year. The average wind speed and the distribution of wind speeds help figure out how much possible energy the turbine can produce. A wind turbine has a power curve that shows how wind speed affects power output. This curve is used to match the turbine to the wind profile at the spot.

Let's say a turbine's power curve shows it can produce 5,000 watt-hours (Wh) each hour when the wind is at its usual speed. If you need 40,000 Wh of energy in a day, you can figure out how long the turbine needs to run by using this equation:

Hours needed = Total energy required ÷ Energy produced per hour

So, in this case:

Hours needed = 40,000 Wh ÷ 5,000 Wh per hour = 8 hours

This means the turbine needs to run for 8 hours to meet the energy demand. To be on the safe side, you might want to account for changes in wind speed by adding a safety margin, so the actual time could be between 7.2 to 9 hours daily.

Turbine Placement

The best place to put a turbine depends on a lot of things—like the direction of the wind, the site's terrain, and environmental concerns. Moving the turbine to a spot where it can catch the most wind, with the least amount of noise and other obstacles, is the goal.

The main things to think about are the wind's direction and speed. Wind turbines should be put in places where the wind blows strongly all the time. To make sure the windmill faces the wind most of the time, the direction of the wind must be considered. For instance, windmills should face westward in places where winds come from that direction most often.

The geography of a site has a big impact on wind flow. Turbines that are put on top of hills or along ridgelines can reach faster wind speeds because they don't have to rub against the ground as much. It is best not to put windmills in valleys, or behind big things like buildings or trees; this can make the wind flow less smoothly and make turbines less efficient. A clear upwind draw, ideally several times the height of the turbine, makes sure that the wind flows smoothly and the turbine works at its best.

It is also important for wind farms to have the right distance between multiple turbines to avoid wake effects, which happen when a turbine changes the wind flow behind it; this makes turbines further downstream less efficient. As a general rule, you should put turbines about 5–8 rotor diameters apart in the direction of the wind that blows the most, and 2–4 rotor diameters apart in the direction that blows the least (Vector Renewables, 2023). This spacing keeps interference to a minimum and increases the total energy capture.

Effects on the environment, like those on wildlife and ecosystems in the area, need to be looked at. Wildlife areas, and the routes that birds and bats take when they migrate, should not have turbines near them.

What kind of foundation is used relies on the soil and the size of the turbine. It needs to offer stable support and spread the weight evenly.

Also, making sure that construction, upkeep, and possible upgrades are easy to get to is very important. Enough access makes installation and continued maintenance go more smoothly.

Homeowners can build and install a wind power system that is both efficient and good for the environment by giving these things careful thought. This will give them reliable renewable energy for their home needs.

Installation Techniques

Wind turbine installations aren't exactly putting together a bookshelf. It's a complex dance of planning, skill, and giant machines to ensure these clean energy giants stand tall and spin safely.

Site Preparation

Setting up the spot is the first step in the installation process. This means getting rid of any obstacles and leveling the ground. Soil tests are done to figure out what kind of foundation is needed, and the spot is marked so that the turbine parts can be put in the right place. Making sure the site is ready for installation assures that everything goes quickly and that the turbine works well once it is in place.

Foundation Construction

The foundation holds the wind turbine firmly to the ground and gives it the support it needs. A shallow or deep foundation may be used, depending on the depth of the base and the type of soil. A big concrete base is made for shallow foundations so the weight is spread out over a lot of space. For home wind turbines, a secure foundation is essential. This is typically achieved by using a sturdy pole or small tower that's firmly anchored in concrete below the ground. Careful installation is key to ensure the stability and long life of your turbine.

Tower Erection

Once the base is set and has dried, the next step is to build the tower. Assembling the tower is done in parts. Depending on how tall the tower is, a Gin Pole or crane is used to lift and bolt each part into place. Because the tower is so tall, the rotor can reach winds that are stronger and more consistent at higher elevations. It is very important to be very precise during this phase to make sure the tower is mounted firmly and vertically. If it isn't, it could affect the performance of the turbine and the structure's strength.

Nacelle and Rotor Installation

The gearbox, engine, and control systems are inside the nacelle—which is then lifted and put on top of the tower. During this step, the nacelle needs to be handled carefully because it is heavy and fragile. After this, the blades of the rotor are connected to the hub. Usually, one blade at a time is put in place, depending on the type and number of blades. Each blade is carefully lifted and bolted to the hub in a planned process. Balance and orientation of the rotor blades have a big effect on how well the turbine works and how long it lasts, so this step is a sensitive one.

Electrical System Integration

Once the mechanical parts are in place, the next step is to connect the electricity system. For this, electric wires are used to connect the generator to the batteries. To make sure that energy is sent efficiently and that electrical problems don't happen, the electrical connections must be made with great care. Control systems are also set up and placed to keep an eye on how the turbine works.

Maintenance and Monitoring

After the wind turbine is installed, it needs to be regularly maintained and watched over to make sure it keeps working well. This includes checking and tightening nuts, lubricating moving parts, and keeping an eye on performance data on a regular basis. Modern monitoring systems can give real-time information about how well the turbine is working, which lets repairs be planned ahead of time and problems be found early.

Safety Measures for Wind Systems

Peace of mind comes with a safe wind turbine system. This means planning carefully, following all the guidelines, and using the best practices at every step—from setting it up to keeping it running smoothly. By prioritizing safety, you'll protect yourself, your investment, and your property.

Site Safety

Proper site safety rules are the building blocks of a safe wind power system. Before the building starts, the site must be carefully checked for any possible dangers—such as rough ground, unstable soil, and power lines nearby. The homeowner should know about safety, what to do in an emergency, how to use personal protective equipment (PPE), and the unique risks that come with building and using a wind turbine.

Construction Safety

Strict safety rules must be followed during the construction part. Lifting must always be done carefully, but it's especially important when lifting heavy things like nacelles and rotor blades. Lifting tools, like cranes, need to be properly maintained and only handled by trained professionals. When working at heights, you have to use fall protection methods, like harnesses. Electrical safety is also very important. All electrical systems must be turned off before work starts and checked using lockout/tagout processes to make sure they don't get turned on by accident.

Operational Safety

Once the wind turbine is up and running, it needs to be constantly checked for safety and maintained. Check for signs of mechanical failure, corrosion, and normal wear and tear through regular checkups. Safety systems, such as brakes and overspeed controls, should be tested regularly, and bolts and moving parts should be lubricated on a regular basis.

Emergency Procedures

Planned responses to emergencies need to be designed. This includes plans for what to do when something breaks, like a machine or an electricity outlet, or when the weather is very bad. Everyone should know how to use fire extinguishers in case of an emergency.

Chapter 10:

Maintaining Your Wind System

Horizontal Axis Turbine and Solar Panel Hybrid System

Keep your home wind turbine running like new! This chapter will guide you through the essential maintenance tasks to get the most out of your system. Regular checkups, keeping things well-oiled, and replacing parts that show wear and tear will prevent problems and extend the life of your turbine. We'll also explore some cool tools that can help you catch issues early and fix them before they become bigger. By following a simple maintenance plan, you'll get the cleanest energy possible from your wind turbine!

Routine Wind Turbine Maintenance

Just like any appliance, your home wind turbine needs regular care to run its best for years to come. Regular checkups help catch small problems before they turn into big ones, keeping your system running smoothly and generating clean energy. These checkups involve looking things over closely, making sure everything is moving as it should, and fine-tuning the system for optimal performance.

Scheduled Inspections

Keeping an eye on your home wind turbine is key to keeping it running great! These checkups, usually done once or twice a year depending on your system and the manufacturer's advice, help catch any issues early. Technicians will look closely at the tower, the nacelle, and the blades for things like wear, rust, or cracks. They'll also make sure everything is nice and tight and there's nothing broken that could stop your turbine from working properly.

Lubrication and Fluid Checks

Keeping things lubricated is like keeping your turbine happy! Just like a car needs oil, your wind turbine needs grease to keep everything moving smoothly. During maintenance, technicians will check and refill the grease in the gearbox, generator, and other moving parts. This helps your turbine last longer and prevents surprise breakdowns.

Rotor Blade Maintenance

To keep them working their best, technicians will clean rotor blades during maintenance to remove dirt, leaves, or even ice that can slow them down. They'll also check the blades carefully for cracks, nicks, or anything else that could be a problem. Keeping the blades clean and in good shape helps your turbine catch the most wind and generate the most clean energy possible.

Electrical System Maintenance

Just like any good appliance, your home wind turbine needs healthy electrical connections to work its best. During maintenance, technicians will check all the wires, connections, and boxes that move the electricity around. They'll also make sure the control panel, any sensors, and anything that talks to the system are working properly. This keeps everything safe and ensures the power your turbine makes gets where it needs to go.

Gearbox and Generator Maintenance

The gearbox and generator are the workhorses of your turbine, and keeping them healthy is important! During maintenance, you will check for things like leaks, strange sounds, or too much shaking in the gearbox. You will also make sure the generator isn't getting too hot, doesn't have any electrical problems, and everything is moving smoothly.

Monitoring Systems

Some systems have special monitors that keep an eye on things like how much power it's making and how hot it's running, all the time. This helps catch any problems early on, so a technician can come out and fix them before they become a bigger issue. It's like having a checkup for your turbine, keeping it healthy and generating clean energy for years to come.

Safety Systems

During maintenance, technicians will check all the systems that help the turbine stop safely in strong winds or bad weather. This includes things like the brakes, the system that turns the turbine to face the wind, and the blades' ability to adjust their angle. Making sure these systems work properly gives you peace of mind, knowing your turbine can shut down safely when it needs to.

Documentation and Recordkeeping

Just like keeping track of your car's service history, keeping good records of your home wind turbine's maintenance is important. Technicians will write down everything they check, clean, or fix during maintenance visits. These notes help them see if there are any problems that keep coming back, so they can address the root cause. Having these records handy is also helpful if you ever need to make a warranty claim on your turbine.

Environmental Considerations

Taking care of your turbine also means taking care of the environment! Technicians will be mindful of things like spills, waste materials, and wildlife when they come to do maintenance. They'll follow best practices to keep things clean and minimize any disruption to the area around your turbine.

Troubleshooting Wind Power Issues

If your home wind turbine isn't producing power like it used to, don't worry! Technicians can troubleshoot the problem to get it back up and running. This means figuring out what's causing the issue, like with any appliance. They'll do a thorough inspection and might use some special tools to pinpoint the exact problem. Once they know what's wrong, they can fix it and get your turbine generating clean energy again.

Common Mechanical Issues

Over time, things like wear and tear, forgetting to grease things up, or parts getting a little out of whack can cause problems in your home wind turbine. For example, strange sounds or shaking could mean there's an issue with the gearbox or the bearings—which are kind of like the gears and wheels that keep everything turning.

Technicians will check these parts for damage, like tiny metal bits in the oil that might mean a gear is wearing down. They can also visually inspect to see if anything is misaligned, and use special tools to fix it if needed. Regular maintenance and replacing parts when they get worn out helps prevent these problems from getting worse and keeps your turbine running smoothly!

Electrical Faults

Electrical problems can sometimes slow your home wind turbine down. This could be due to frayed wires, worn-out parts inside the generator, or loose connections. Technicians will use special tools to check the electrical circuits and make sure everything is connected properly and insulated well.

They'll also look at the parts inside the generator that make the electricity, like the brushes and slip rings, to see if they're worn out. If something is getting too hot, they might use a special camera to see if there's an electrical issue that needs fixing. By catching these problems early, they can keep your turbine generating clean energy efficiently.

Rotor Blade Issues

The blades on your home wind turbine face a lot throughout the year, from sun and wind to rain and even ice! Over time, this can cause cracks, nicks, or ice building up on the blades, which can make them less efficient at catching wind. Technicians will give the blades a close inspection, sometimes using a remote-controlled flying machine (like a drone) to get a good look at everything. They might also use a special camera that can see heat, to check for hidden cracks or damage inside the blades. If they find any problems, they'll fix the blades or replace them if needed.

Gearbox Problems

Over time, if things aren't lubricated properly, the gears and bearings can become worn—all of which can slow the turbine down. Technicians can check the oil to see if it's dirty or needs replacing, and they can also use special tools to listen for any unusual sounds or vibrations that might signal problems with the gears or bearings.

Yaw and Pitch System Failures

To catch the most wind possible, your home wind turbine needs to be able to turn with the changing wind direction and adjust the angle of its blades. These adjustments are made by the yaw and pitch systems. If these systems aren't working properly, your turbine might not be catching as much wind as it could, or it could be working a little too hard. Technicians will check the motors and gears that turn the turbine, and make sure the parts that adjust the blades are moving smoothly. They'll also keep everything well-lubricated and make sure everything is calibrated correctly.

Predictive Maintenance

Imagine your home wind turbine being super smart and letting you know if something might be about to go wrong. That's kind of what predictive maintenance is like. By constantly monitoring how the turbine is working and using clever computer programs, technicians can predict problems before they happen. This means they can fix things during regular maintenance visits, instead of waiting for a breakdown.

Expanding and Upgrading Wind Installations

Thinking about getting even more clean energy from your home wind turbine? Adding on to your existing system can be a great option! Before you jump in, it's important to do some planning to make sure it all works together smoothly. This might involve considering how much extra power you need, what kind of new equipment might be best, and if there have been any advancements in wind turbine technology since you first installed yours. Here's a breakdown of the key steps involved.

Step 1: Checking Up on Your Current System

Check how your current turbine is doing. This involves looking at how much power it's making, how often it's been needing maintenance, and any historical data you have. Are there any parts that seem to be underperforming? Understanding any limitations will help guide your upgrade decisions.

Step 2: How Much Extra Power Do You Need?

Take a look at your past energy usage and how much you expect it to grow in the future. Are you planning on getting an electric car or adding new appliances? This will help determine how much extra power you'll need from your expanded system.

Step 3: Choosing the Right New Turbine

Newer wind turbines are much better than older models. They're more efficient, reliable, and can generate more power even in lower wind speeds. Think about advances in blade design and materials that can make a big difference. Upgrading to a larger, more efficient turbine might be all you need to meet your energy requirements.

Step 4: Repowering Your Existing Site

Instead of a whole new setup, you can also consider replacing your current turbine with a newer, more powerful one. This could involve using the same foundation or building a new tower entirely. This can be a great way to get a big boost in energy production and lower your maintenance costs in the long run.

Step 5: Upgrading Your Electrical Grid

If you're planning on generating more power, you might need to make some adjustments to your electrical system. This could involve strengthening your connection to your power grid or even adding battery storage solutions. The goal is to make sure your system can handle the extra electricity efficiently.

Step 6: Keeping Things Green and Legal

Before expanding, it's important to consider any environmental impacts. Are there any wildlife or habitat concerns in your area? There might be some assessments you need to complete.

Step 7: Planning Your Budget

Expanding your wind system will involve some costs. Do some research and consider any government incentives or subsidies that might be available to help offset those costs.

Step 8: Taking Care of Your Upgraded System

Just like your current turbine, regular maintenance and monitoring are key to keeping your expanded system running smoothly. New technology might involve advanced monitoring systems that can help predict and address any problems quickly. Regular checkups and maintenance will ensure you get the most out of your investment for years to come!

Chapter 11:

Battery Storage Options

Batteries and Control System

Imagine if your home wind turbine could keep making clean energy even when the wind isn't blowing as hard. That's the magic of battery storage! This chapter will explore how batteries can be a backup for your turbine, storing any extra electricity it generates during the day so you can use it at night or on calmer days.

We'll look at different types of home battery systems, like lithium-ion (similar to the battery in your phone), lead-acid (like in a car battery), and flow batteries (a little more complex, but still cool). Each has its pros and cons, but all can help you become more energy independent and rely on clean power for longer.

By understanding battery storage, you can see if it's a good fit for your home wind turbine system and how it can help you maximize your clean energy production!

Types of Batteries

Keeping extra energy in batteries for later use is an important part of green energy systems that make them more reliable. There are many kinds of batteries, but the most popular ones are lead-acid, lithium-ion, and flow batteries. Each has its own features and uses.

Lead-Acid Batteries

Lead-acid batteries have been around for a long time, and you might even recognize them from your car! These batteries are a well-known choice for home wind turbine systems because they're:

- **Advantages**
    - **Reliable:** They've been around for a long time, so you know they're dependable for storing your clean energy.
    - **Affordable:** Lead-acid batteries are generally less expensive than some other battery options.

They work by submerging lead plates in an acid solution, which might sound a bit complex, but they're a proven technology for home energy storage. However, it's important to note that they don't last quite as long as some other options:

- **Disadvantages**

- Size and weight: Lead-acid batteries can be pretty bulky and heavy. This might not be an issue if you have a lot of space, but it's something to keep in mind, especially for smaller homes.

- **Shorter lifespan:** They generally don't last as long as some other battery options. Think of it like needing to replace the battery in your car more often. With proper care, they can still provide good service, but it's something to factor in.

- **Less storage capacity:** For their size and weight, lead-acid batteries don't store quite as much clean energy as some newer battery technologies. This means you might need more batteries to store the same amount of power.

Lead-acid batteries are a familiar sight in a few places. They're sometimes used in those backup battery systems that keep your computer running during a power outage. You might also find them being used alongside solar panels or other renewable energy sources to store extra clean energy.

Lithium-Ion Batteries

Lithium-ion batteries are like the rockstars of the battery world! They're becoming increasingly popular for home wind turbine systems because they offer some major benefits:

- **Advantages**

  - **More power in a smaller package:** These batteries can store a lot more clean energy for their size and weight compared to lead-acid batteries. This means you might need fewer batteries to store the same amount of power, saving you space.

- **They last longer:** Lithium-ion batteries can typically go through many more charge and discharge cycles than lead-acid batteries. Think of it like needing to replace your phone battery much less often.

- **Less energy waste:** Lithium-ion batteries are very efficient, so you lose less clean energy when you store it and then use it later.

However, there are also a couple of things to keep in mind:

- **Disadvantages**

  - **Higher upfront cost:** Lithium-ion batteries are generally more expensive than lead-acid batteries.

  - **Safety first:** They can sometimes overheat, so they require special systems to keep them safe.

Despite the higher cost, lithium-ion batteries might be a good choice for your home wind turbine system if you have the space for lead-acid batteries but want more storage capacity or a longer lifespan.

Flow Batteries

Flow batteries are a new technology that's still being developed for home use. They work a little differently than other battery options by storing energy in liquid solutions that flow through a special chamber. Here's a quick rundown of their pros and cons:

- **Advantages**

  - **Super scalable:** If you ever need more storage capacity in the future, you can easily add more tanks to your flow battery system. This makes them a flexible option.

- **Long lasting:** Flow batteries can last through many charging and discharging cycles, so you might not need to replace them for a long time.
- **Safety first (like lead-acid):** Flow batteries are less likely to overheat compared to some other options.

However, there are also a few things to keep in mind:

- **Disadvantages**
  - **More complex setup:** Flow batteries involve more parts, like pumps and tanks, which can make them more complicated to install and maintain.
  - **Larger footprint:** For the same amount of storage capacity, flow batteries currently take up more space than some other battery options.

Flow batteries are a promising technology for home wind turbine systems, but they might not be the most common choice just yet. As the technology develops, they could become a more viable option for homeowners in the future.

Battery Efficiency and Lifespan

The performance and cost viability of energy storage systems depend on how well and how long the batteries last. These factors affect how well a battery can store and release energy over time, as well as how long it will work before it needs to be replaced.

Battery Efficiency

The amount of energy lost during the charge and discharge processes is measured by battery efficiency, which is also sometimes called round-trip efficiency. It is given as a proportion of the energy that was put in and is taken back during discharge. During these steps, high-efficiency batteries lose less energy, which makes them cheaper and better for the earth.

Factors influencing efficiency include:

1. **Internal resistance:** When batteries are charged and discharged, they lose less energy as heat when their internal resistance is low. This makes things work better and makes sure there is more energy accessible.

2. **Temperature:** Temperature has a big effect on how well batteries work. Most batteries work best when the temperature is mild and, if the temperature changes, the battery may not work as well. When temperatures are low, chemical processes slow down, which makes the system less efficient. When temperatures are high, internal resistance goes up.

3. **Charge/Discharge rates:** Charging and draining quickly can make the battery less efficient because it creates more heat and internal resistance. The best efficiency is reached at modest rates that are in line with how the battery was designed.

Let's take a look at the efficiency of different battery types (Fitton, 2024):

- **Lead-Acid batteries:** For lead-acid batteries, the round-trip efficiency is usually lower, between 70 and 80%. Energy is lost because of internal resistance and the heat that is made when chemicals respond.

- **Lithium-Ion batteries:** Lithium-ion batteries are known for being very efficient. They usually get 90% to 95% energy round-trip. These high yields are due in part to their low internal resistance and advanced charge control systems.
- **Flow Batteries:** Most of the time, flow batteries are only 65% to 85% efficient. Flow batteries lose energy because they need to pump energy to move the liquid around and the cell stacks aren't very good at converting energy.

Battery Lifespan

Imagine a battery like a well. You can draw water (energy) out of it, but eventually, the well runs low. That's kind of what happens with battery lifespan. It's how long a battery can work well before it needs a recharge or, in this case, a replacement. We usually measure this in cycles—how many times you fill and empty the well—or just years.

So, what affects how long this well keeps giving? A few things. One is how much water you take out each time—Depth of Discharge (DOD). If you take out a lot each round, it refills slower and might run dry faster. Batteries like to work within a certain range for best results.

Temperature is another factor, just like with most things. Extreme heat can wear the well down quicker, while freezing cold can make it sluggish and even damage it. Another thing to consider is the cycle life, which is basically how many times you can fill and empty the well before it gets significantly shallower. A higher cycle life means the well holds up for longer.

Finally, taking good care of your battery, like following charging instructions, is like maintaining the well. It helps it last as long as possible.

Lifespan of Different Battery Types

The average life of a lead-acid battery is 500 cycles at 50% DoD, which is significantly less than that of other types of batteries (Foxtron, 2021). They need to be changed more often than other types of batteries, but if you take care of them, they can last longer.

But lithium-ion batteries last a lot longer; at an 80% DoD, they can often last over 2,000 rounds (天空 [Sky], 2024). Some very high-tech types can even go through 5,000 cycles. They are a popular choice for many uses because they have a high energy density and don't need much upkeep.

Flow batteries are made to last a long time; they can often go over 10,000 rounds (Christian M., 2023). Because of how they're made, the electrolyte can be easily replaced, which essentially recharges the battery and extends its useful life. So, flow batteries are great for long-term uses where longevity is important.

Maximizing Battery Efficiency and Lifespan

Just like Goldilocks and the porridge, batteries like things to be "just right." That means keeping them in a temperature that's not too hot or too cold (check the manual for the sweet spot). Also, try not to drain them all the way down each time (Depth of Discharge)—follow the recommended limits. Think of it like not squeezing the toothpaste tube all the way from the bottom every time!

Regular checkups are good for your battery, too, just like going to the dentist is good for your teeth. This might involve balancing the cells inside the battery to keep things running smoothly. Fancy gadgets called Battery Management Systems (BMS) can even be your battery's personal assistant, keeping an eye on temperature, charging cycles, and other things. This helps your battery work efficiently and last a lot longer, saving you money and frustration in the long run.

Storage Solutions and Installation

Storage options for sustainable energy sources, like solar and wind power, make sure that energy is reliable and used efficiently. It's just as important to install these storage systems correctly if you want them to work well and last a long time. Here's a breakdown of what goes into installing a storage system to keep your renewable energy flowing smoothly.

Making Sure Your Home Is Ready

First things first, you need to see if you have enough room for the system. You'll also want to take a look at your electrical wiring and the climate around your house. Just like a comfy temperature keeps you happy, the right environment keeps your battery happy too! Make sure it has good ventilation and protection from the elements.

Matching Your Needs

Pick the perfect battery type, size, and setup for your energy use. Think of it like choosing the right size cooler for your next picnic! For some batteries, you'll design a special system (BMS) to keep an eye on each battery cell, just like making sure all your food stays nice and cold. You'll also connect your storage system to your existing solar panels or wind turbine and the main grid (power company), making sure everything works together seamlessly.

Putting It All Together

This is where things get hands-on! You'll place the batteries, inverters (which turn the stored energy into usable power), and other equipment in their designated spots. Depending on the battery type, you might need to set up special ventilation or temperature control systems. Safety first! Follow all the manufacturer's instructions and safety standards to make sure everything is installed properly.

Connecting to the Grid

You'll carefully connect the batteries to the inverters and, then, to your home's electrical system. Think of it like connecting the cooler to your picnic basket! This ensures all the connections are secure to prevent any electrical problems. If you're also connecting to the main grid, you'll need to work with your utility company to make sure everything integrates smoothly.

Making Sure It Works

Once everything is installed, you'll put the system through its paces! Check all the connections, make sure the battery management system is working properly, and see how the system handles different power demands. It's like giving your new cooler a test run before you fill it with drinks!

Keeping it Running Smoothly

Just like any appliance, your battery storage system needs some tender loving care (TLC) to keep it working well for a long time. Set up a schedule for regular checkups and maintenance, keeping an eye on things like battery health and performance. Advanced monitoring systems will also be installed to give you real-time data on how your battery is doing, so you can catch any potential issues before they become a problem.

By following these steps, you can ensure your renewable energy storage system is installed properly, works efficiently, and keeps your clean power flowing for years to come!

Chapter 12:

Inverters and Converters

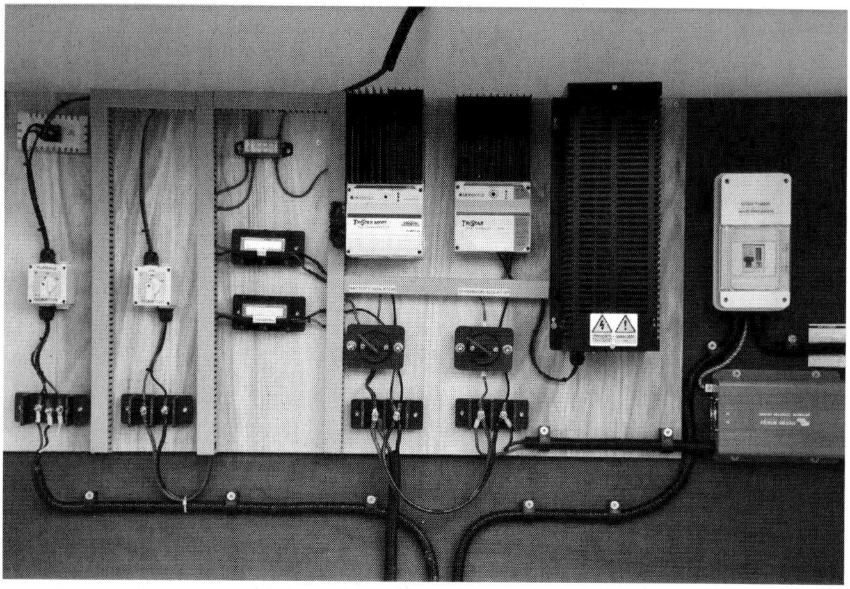

System Layout

Sunshine and wind are fantastic energy sources, but they need a little help to run your appliances. This chapter unlocks the secrets of inverters and converters: the behind-the-scenes heroes of renewable energy systems. We'll explore how these devices transform the energy from your solar panels or wind turbine.

Converters are like traffic controllers, managing the voltage levels of the electricity to ensure smooth and efficient flow. They fine-tune the power to match your needs and prevent any hiccups.

Understanding these key players is essential for getting the most out of your renewable energy system. This chapter will be your guide; we'll show you how to choose the right inverter and converter, set them up for optimal performance, and keep them running smoothly. By mastering these essential components, you'll ensure your renewable energy system delivers clean power to your home efficiently.

Functions and Selection Criteria of Inverters and Converters

For renewable energy systems to work, inverters and converters must be in place to efficiently handle and transform electrical energy. Mostly, they manage voltage levels and transform DC power from sources like solar panels or batteries into AC power that homes may utilize. To get the best system performance, you need to know how the inverters and converters work and think about a number of factors when choosing them.

Functions of Inverters

DC electricity from solar panels or batteries is turned into AC electricity by inverters. Most home products use AC electricity, and extra AC electricity can be sent back to the power grid. There are various kinds of inverters, and each one is used for a particular task:

- **String inverters**: These are the types that are most often used in solar panels. They link a string of solar panels together and change the DC output to AC. These panels are inexpensive and can be used in places that get a lot of sun and not much shade.

- **Microinverters**: These convert direct current (DC) to alternating current (AC) on the level of the individual solar panels. Since each panel can work on its own, they are perfect for setups that have problems with shading or complicated roof designs.

- **Central inverters**: Central inverters take the DC output from many sets of panels and turn it into AC. They are used in large solar farms. They're very efficient and don't cost much for big setups.

- **Hybrid inverters**: These can handle power from the grid, batteries, and solar cells. They're used in systems that need to combine solar power with battery storage and power from the grid to make sure there is always electricity.

Functions of Converters

Different types of DC can be changed and voltage values can be changed. For the efficient management of energy storage and the guarantee of component compatibility, they are indispensable:

- **DC-DC Converters**: These control the voltage levels from the batteries or solar panels, making sure the voltage is right for the load or the generator. They help the solar panels work better by using maximum power point tracking (MPPT), which changes the modules' electrical working point to get the most power out of them.

- **AC-DC Converters**: These take AC power off the grid, convert it into DC power, and put it into batteries. When solar output is low, these systems rely on grid electricity to keep the batteries charged.

Guide to Inverter and Converter Selection

Equipping your renewable energy system with the perfect inverter and converter is crucial for optimal performance. Here's a breakdown of key factors to consider.

Power Up for Your Needs

Just like any good team player, your inverter's capacity should match your peak power requirements. Think beyond average usage and consider situations with high energy demands. If your peak load is 5 kW, a slightly higher-rated inverter ensures smooth sailing.

Minimize Energy Loss

Look for inverters and converters boasting efficiency ratings above 95%. These champions minimize energy loss during conversion, squeezing the most juice out of your renewable sources.

Waveform Matters

Inverters come in different waveform flavors. Pure sine wave inverters provide the cleanest power, mimicking the grid and keeping your sensitive electronics happy. Modified sine wave options are more budget-friendly but might cause hiccups with some devices.

Compatible Connections

For systems with battery backup, ensure your inverter and converter play nicely with your battery type and voltage. This creates a harmonious charging and discharging cycle, maximizing battery life.

Grid Integration

For grid-tie systems, compliance with local regulations is essential. Features like anti-islanding protection, which prevents power backfeed during outages, are key for safety and grid harmony.

Future-Proofing

Thinking of expanding your system in the future? Opt for modular inverters and converters. These allow for easy addition of more solar panels or batteries without a system overhaul.

Durability and Warranty

Consider the environment where your system resides. Choose inverters and converters that can handle temperature swings, humidity, and other local weather challenges. Look for warranties of 10 years or more for peace of mind and long-term support.

Keeping an Eye on Performance

Advanced inverters and converters offer built-in monitoring capabilities. Track system performance in real-time, which allows you to identify and address any issues promptly. Options like Wi-Fi or Ethernet for remote monitoring add another layer of convenience to system management.

By considering these selection criteria, you'll be well on your way to choosing the perfect inverter and converter duo for your renewable energy system. Remember, a well-matched team is a happy (and efficient) team!

Integration of Inverters and Converters With Solar and Wind Systems

While it's true that wind turbines and solar panels can collect wind and solar energy, it's the inverters and converters that turn that energy into something practical. Adding these important parts to your renewable energy system is like adding the secret sauce. Get the most out of your system in terms of power and reliability by making sure all the elements operate together seamlessly through careful design and configuration.

Integration With Solar Systems

Solar cells generate electricity, but not the kind that powers our homes and businesses—it's *direct current* (DC). To bridge this gap, we rely on inverters: the transformers of the solar world. These workhorses come in different configurations, each suited to specific needs.

For setups with many solar panels linked together, string inverters are the go-to choice. They efficiently convert the combined DC output from a string of panels into usable *alternating current* (AC). These cost-effective inverters are ideal for areas with minimal shade.

For sites with shading or complex roof designs, microinverters offer a unique solution. Each solar panel has its own built-in microinverter, converting DC to AC power at the panel level. This allows each panel to function independently, maximizing energy production even in shaded areas.

For vast solar farms, central inverters take center stage. They handle the heavy lifting, converting the DC output from multiple strings of panels into AC power. These are highly efficient for large-scale systems.

But the story doesn't end there. DC-DC converters play a crucial role in ensuring optimal performance. They adjust the voltage level from the solar panels to match the specific input requirements of the inverters. Additionally, maximum power point tracking (MPPT) technology is often integrated into these converters. MPPT acts like a smart manager, constantly adjusting the operating points of the panels to squeeze the most power out of them.

For systems that incorporate battery storage, hybrid inverters become the brains of the operation. They manage the flow of power from solar panels, batteries, and even the grid—ensuring a constant and reliable energy supply. These intelligent inverters determine when to charge and discharge batteries, optimizing energy use and storage based on real-time production and demand.

Integration With Wind Systems

Wind turbines are great at collecting energy from the wind, but sometimes they make crazy amounts of power. Converters are the unsung stars that make sure power is delivered smoothly and can be used with the grid.

DC-DC converters are the most important part of wind turbines that make DC energy. They change the entering DC voltage to exactly what the inverter needs, which is similar to how voltage regulators work. This makes sure that the power exchange is stable and effective, even if the wind speed changes.

The process is a little more complicated for wind turbines that make AC energy. Changers from AC to DC to AC are used. First, these multitaskers change the variable AC output from the engine to DC by rectifying it. Then, they change the stable DC power back to clean AC power, which is what converters do. This two-step process makes sure that the end output works with the frequency and voltage of the grid.

When we talk about the grid, grid-tie inverters are very important to wind energy sources. These inverters constantly match the frequency and voltage of the power from the rotor to what the grid needs. They do this by acting like synchronizers. This smooth integration makes sure that the power from the wind turbine goes into the grid without any problems. More advanced inverters go one step further by changing their output based on how the grid is doing in real-time. This makes the integration process even better.

Like solar systems, wind systems can benefit from having batteries built in. In these kinds of setups, hybrid inverters are like the brains of the business. They make sure that power moves smoothly between the batteries, the wind turbine, and the power grid. These smart inverters improve the general reliability and efficiency of the wind energy system by looking at patterns of generation and usage. They do this by optimizing battery storage and release.

Lastly, power from wind turbines may have fluctuations and harmonics that lower the quality of the power, as a whole. Power dryers will take care of you! These systems, which usually have filters and processors, clean and smooth the power. They get rid of these unwanted changes and waves, making sure that the final power output is clean and up to the grid's strict standards.

Combined Solar and Wind Systems

Combine the force of the wind with the radiance of the sun, and you have a system that harnesses the full potential of both sources of energy. But how do you keep these two sources of power in check and make sure the flow is smooth and reliable? The hybrid inverter tops here.

These flexible parts are what make solar and wind systems work together. They handle the changing amounts of energy from both sources without any problems, keeping things stable and meeting your power needs. That's not all hybrid inverters can do. They also work well with battery storage, which gives you even more options and peace of mind. When batteries are used, extra energy can be saved for later use, which makes the system even more efficient.

Of course, careful planning is needed to make a combination system work well. The key is to make sure that the power from the sun and the wind are balanced. To do this, you need to choose inverters and converters that work well together and can handle the changes from both sources. Also, the battery storage system needs to be the right size to handle the flow of all the energy.

The key to managing a solar and wind system together is to use advanced monitoring and control tools. These systems keep an eye on things and send real-time information about how much energy is being made, used, and stored. Since this information is easy to get to, you can control and tweak your whole system to run at its best.

People who connect their power system to the grid must make sure they follow the rules for that grid. These rules are in place to keep things safe and to keep the grid stable. By following these tips, you can make sure that your solar and wind system works safely and without any problems with the grid.

Installation and Maintenance of Inverters and Converters

Think of inverters and converters like the translators of your renewable energy system! They take the electricity language spoken by your solar panels or wind turbine and turn it into the language your home understands. Proper installation makes sure everything is wired up correctly, and regular maintenance catches any problems before they happen. This keeps your whole system running smoothly and lets you enjoy clean energy for a long time!

Installation of Inverters

Site Evaluation and Planning

Before installing the inverter, you should look at the site to find the best spot for it. To keep it from getting too hot, you should put it somewhere cool, dry, and well-ventilated. Important things to think about are how easy it is to maintain and how well it's protected against the weather.

Setting Up the Inverter

Place the inverter on a stable surface, like a wall or a rack made just for it. Make sure it's securely attached to stop vibrations and possible harm. The place where the inverter is mounted should leave enough space around it for air to flow and heat to escape.

Electrical Connections

Make sure the polarity is right when you connect the DC input from the solar panels or wind blades to the inverter's DC terminals. Use cables and connections that are rated to handle the expected voltage and current. Next, connect the AC output to the home or grid link, making sure to follow the safety rules and electrical codes in your area. It is important to properly ground the inverter to protect it from electrical problems.

Configuration

Set up the inverter so that it works with the system. This includes setting the voltage and frequency factors. A lot of inverters have digital displays or software tools that let you set them up. After setting up the system, it needs to be put into service by running some basic tests to make sure all the connections are safe and the inverter is working correctly. During the first run, keep an eye on the inverter to see if anything seems off.

Installation of Converters

Site Selection and Preparation

Converters should also be put in a cool, dry, and well-ventilated area, just like inverters. Make sure the spot is safe from the elements and easy to get to for regular maintenance.

Mounting and Securing

You can put the adapter on a stable surface or inside a case that protects it physically. Firmly fasten the item to avoid vibrations or movement that can cause mechanical stress or electrical problems.

Wiring and Connections

Connect the input and output terminals carefully, making sure to follow the right voltage and current values. Connect the DC-DC converter's input to a green energy source and its output to a load or a battery bank. To keep electrical resistance and heat from building up, use high-quality connectors and make sure they are tight and secure.

System Integration and Testing

Add the converter to the energy system as a whole, making sure it is set up to meet the needs of the load or storage system. Turn on the device and check for stable voltage and current outputs to test the converter. Keep an eye on the system for a while to make sure it works correctly under different loads.

Maintenance of Inverters and Converters

Just like any hardworking appliance, your inverters and converters need regular checkups to keep them running smoothly and efficiently. Here's a breakdown of how to maintain them:

1. Start with a thorough visual inspection. Look for any signs of wear and tear, such as loose connections, rust, or cracks in the casing. Check the fans and heat sinks for dust buildup, which can hinder proper cooling.

2. In addition to regular inspections, keeping your inverters and converters clean is crucial. Wipe down the surfaces periodically and remove dust from the air grilles and fans. Make sure the installation site remains free of clutter or obstructions that could block airflow and lead to overheating.

3. Move on to electrical testing to ensure everything is functioning within normal parameters. Use a multimeter or other diagnostic

tools to check voltage, current, and resistance levels. Additionally, test the safety features, like circuit breakers and fuses, to verify they'll activate if a fault occurs.

4. Similar to your phone, inverters and converters might require firmware updates from the manufacturer. These updates often include performance improvements, bug fixes, and new features that can enhance system efficiency and reliability.

5. Don't rely solely on scheduled checkups. Utilize built-in monitoring systems or external diagnostic tools to continuously monitor the performance of your inverters and converters. Track key metrics, such as temperature, energy output, and efficiency. Analyze this data to identify any trends or anomalies that could indicate potential issues, allowing you to address them before they escalate.

6. Finally, schedule preventive maintenance tasks based on the manufacturer's recommendations and your system's operational data. This may involve recalibrating settings, tightening electrical connections, and replacing worn components. Taking proactive steps like these helps prevent unexpected failures and extends the lifespan of your valuable equipment.

Chapter 13:

Monitoring and Control Systems

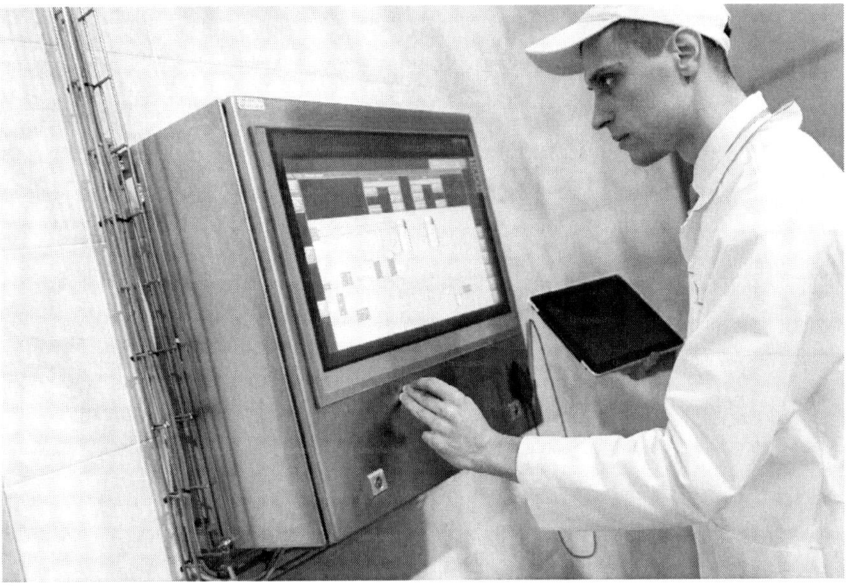

One Form of Control Panel

In this chapter, we will examine the control and monitoring systems, which are the central nervous systems of renewable energy facilities. They are very important because they keep an eye on the system and make changes to ensure it works well for years to come.

Take a look at a network of sensors that monitor everything from the health of the batteries to how much energy is being made. A control center receives this data in real-time, much like a mission control center for renewable energy. This is where the system does its heavy lifting, analyzing data to make informed judgments like intelligent power distribution or automated problem-solving.

Importance of Monitoring Systems

Monitoring tools are necessary for renewable energy installations to work well and reliably. One of their main jobs is to track success in real time. They are always gathering information on things like energy production, system efficiency, and operational state. This information is very helpful for understanding how wind and solar power systems work on a daily basis. This real-time data helps find performance differences from what was predicted, so quick fixes can be made to keep energy output at its best.

Fault Detection and Diagnostics

Diagnostics and finding faults depend on monitoring tools in a big way. By looking at data trends and finding outliers, they can find problems early—like solar panels giving off strange temperature readings or wind turbines shaking in strange ways. Early detection lets help be sent right away, which diminishes downtime and keeps expensive fixes from having to be made.

Enhanced Safety

Another important thing is safety. Monitoring systems check the working state of important parts, all the time, sounding alarms and shutting down automatically if they find dangerous situations, like overheating or electrical problems. This makes sure that all of the parts work within safe limits, which keeps accidents from happening and keeps people and tools safe.

Optimized Maintenance

Renewable energy systems need to be properly maintained in order to last a long time and work reliably. Monitoring systems make predictive maintenance possible by giving thorough information about how different parts are working and what their conditions are. This method, called *condition-based maintenance*, makes the best use of resources, lowers upkeep costs, and increases the life of equipment.

Energy Management and Efficiency

Monitoring systems keep track of how much energy is made, used, and stored. This helps maintain the balance between supply and demand. They control the flow of energy between different sources in mixed systems, which makes the whole thing more efficient. Monitoring data can also show where energy is being wasted, which lets changes be made more precisely.

Making Decisions Based on Data

Monitoring systems collect data that helps operators make decisions based on facts. They can look at both past and real-time data to learn more, spot trends, and make smart choices about whether to upgrade or change the system. In turn, this makes green energy projects more useful and profitable.

Improved Reliability and Uptime

Overall, monitoring systems make things more reliable and are available all the time. They detect issues and identify problems early on, which cuts downtime and makes sure there is a steady flow of power. They also make it easier to connect to smart grids by giving real-time data and control for better energy distribution, load balancing, and grid safety.

Available Technologies and Tools for Monitoring Systems

These state-of-the-art systems collect information from wind and solar power systems. The data is then analyzed and presented in a way that enables proactive management and guarantees efficiency, safety, and peak performance.

Sensors and Data Collection Devices

Utilizing these sensors and devices can help you keep track of important data:

- **Temperature sensors**: Check the temperature of the solar panels and generators to keep them from getting too hot.
- **Current and voltage sensors**: Assess how much electricity solar panels and wind turbines produce.
- **Anemometers**: For the best performance of wind turbines, measure the speed and direction of the wind.
- **Pyranometers**: Check the sun irradiance to see how well the solar panels work.

Data Loggers and Acquisition Systems

Data loggers take measurements all the time and store them so that they can be analyzed later. They work with many sensors and connection protocols, and more advanced models do preliminary data processing to cut down on the amount of data that needs to be sent.

Data Management and Analytics Platforms

These software programs get data from tracking systems, store it, and look at it. Some important features are:

- **Data visualization:** These dashboards present historical and real-time data through graphs, charts, and heat maps.
- **Analytics and reporting:** Machine learning and algorithms find patterns, predict failures, and make processes run more smoothly.
- **Integration capabilities:** Connect to business tools to get a full picture of how things are running.

Predictive Maintenance Tools

Predictive maintenance uses data from monitoring to guess when equipment will break down, which allows repairs to be made before anything happens. Thanks to cutting-edge analytics and machine learning, this increases equipment lifespan, cuts costs, and decreases downtime.

Cybersecurity Solutions

Cybersecurity is vital for renewable energy systems since they are interdependent. Some such solutions are:

- **Intrusion Detection Systems (IDS) and firewalls:** IDS and firewalls prevent inappropriate access and identify security breaches.
- **Encryption:** This protects data transmission against spying and manipulation.

- **Control access:** Using role-based controls and multi-factor authentication, it limits system access to authorized personnel.

When used together, these resources make solar and wind power systems more secure, reliable, and efficient. This is primarily for commercial systems, but can be incorporated into home systems.

Installation and Use

Making sure your monitoring systems work right is important for ensuring your green energy system works well and lasts a long time. To get everything just right, you need to carefully plan, carry out, and keep an eye on things:

1. ou should first look over your site. Take a look at the layout and decide where the best spots are for your sensors. Also, make sure there is good coverage and easy access for upkeep. Don't forget to think about things like temperature, humidity, and how much the system will be exposed to the weather.

2. Next, be very careful when picking out your gear. Choose the correct sensors, data loggers, and transmission tools for your installation based on its needs, size, and the weather it will be exposed to. Make sure that everything works with the way things are set up now and that it can be improved if needed in the future.

3. When you put the monitors and data loggers in place, where you put them is very important. To get correct readings on temperature, voltage, current, wind speed, and solar irradiance—sensors should be placed in a planned-out way. Secure them properly to keep them from moving or getting damaged. Put the data loggers somewhere central that is easy to

get to, make sure that all of the links are safe, and follow all safety rules and standards.

4. Another important step is to set up a reliable contact network. Choose between wired and wireless contact, based on your needs. High-speed links that are stable can be made with Ethernet or fiber optics, but they may need more infrastructure. Cellular networks, Wi-Fi, and radio frequency are all options for places that are far away. Make sure you have a strong signal and that your info is safe.

5. Last but not least, set up and test your system. Set up your data acquisition and loggers to take measurements at regular times, and make sure that your sensors are calibrated to take into account any practical or environmental factors. Fine-tune everything for the best results with software tools or digital interfaces. Your green energy installation will continue to run smoothly and efficiently as long as you keep the software and firmware updated and manage it.

Maintaining a Monitoring System

Think of the monitoring system as an early warning system. By analyzing data patterns, the system can flag potential problems, allowing for prompt interventions and preventive maintenance. Advanced tools can even predict failures before they happen, minimizing downtime.

But just like any workhorse, your monitoring system needs regular care. Routine maintenance includes inspections, cleaning, and calibration of sensors. Don't forget to update software and firmware for optimal performance. Keeping detailed records of maintenance and addressing any issues promptly ensures your system remains reliable for years to come.

Data is king, and managing it effectively unlocks the system's true potential. Secure storage and regular backups safeguard your valuable information. Utilize analytical tools to interpret data and generate reports with actionable insights. Automated reporting features can further streamline the process—aiding operational strategies, maintenance planning, and regulatory compliance.

Chapter 14:

Benefits of Hybrid Systems

Hybrid System With Roof-Mounted Panels Depiction

Hybrid power systems combine the best of both worlds: solar and wind energy. This chapter dives into the reasons why this combo is a champion for reliability, efficiency, and sustainability. Hybrid systems use multiple sources, meaning they're not at the mercy of the weather. When the sun isn't shining, the wind can pick up the slack. This flexibility keeps the energy flowing and reduces dependence on a single, unpredictable source.

But that's not all! Hybrid systems are designed to be eco-friendly by maximizing renewable resources and minimizing environmental impact. They also boost energy security by providing a more reliable and independent power source. With all these advantages, it's no wonder hybrid systems are becoming a star player in the game of sustainable energy solutions.

Combining Solar and Wind Power

Using solar and wind power together in hybrid systems makes energy more reliable, efficient, and environmentally friendly by taking advantage of the fact that these two sustainable energy sources work well together. Wind power can be made at any time of the day, but it is usually better at night and in the winter. Solar power is most powerful during the day. This makes the energy supply more stable and constant, since each source makes up for the intermittent nature of the other.

Hybrid systems make energy much more reliable—especially in remote or off-grid areas—because they provide steady power even on cloudy or calm days. They also make the best use of resources, which means they produce the most energy all year long. Hybrid systems can use both sunshine and wind at the same time, which means they can produce more energy than single-source systems in places with lots of both.

Combining solar and wind power has less of an effect on the environment because it reduces the need for fossil fuels and greenhouse gas pollution. Because they share infrastructure, like inverters and transmission lines, hybrid systems also take up less space than separate installations. This is another reason they're better for the environment.

Advantages and Challenges of Hybrid Power Systems

Hybrid systems are like the ultimate energy team-up, combining solar and wind power. They're awesome for sustainable energy, but there can be a few bumps in the road. Let's explore both sides to see why hybrid power is still a champion!

Advantages

Mixed methods are great because they don't just use one source. Solar and wind power work together, which is like having a backup plan for your energy needs. In other words, the wind can make up for the sun's lack of brightness. No matter what the weather is like, it's all about keeping the energy flowing easily. This level of dependability changes everything, especially in remote places where the grid might not work as well as it should.

To put it another way, hybrid systems are like the best resource controllers. They get more energy from both the sun and the wind than either source could provide on its own all year. If you were in a place with a lot of sun and strong winds, a combined system would work great because it would use both resources to make more energy. Plus, using both sources well means better results and getting the most out of your investment in green energy.

There's more, though! The cost of making one unit of energy goes down when solar and wind power are used together. You don't need those pricey backup systems, either, because they're so stable. This saves you even more money in the long run.

Challenges

Hybrid systems sound amazing, right? Sunshine and wind power working together—what's not to love? Well, there are a few hurdles to jump before they're everywhere.

First, think of a complex juggling act. Hybrid systems combine solar panels, wind turbines, and even the power grid—which requires some serious tech skills to manage. Fancy control systems and smart grids are needed to keep everything balanced and working smoothly. This can be a barrier for some, as it requires expertise and can mean more maintenance down the line.

Then there's the upfront cost. All that fancy equipment doesn't come cheap. While hybrid systems save money in the long run, the initial investment can be a big gulp. Securing funding and incentives can be crucial to getting these projects off the ground.

Another challenge is that sunshine and wind aren't always reliable. To keep the lights on all the time, hybrid systems often need batteries to store extra energy. But batteries can be expensive, and the technology isn't quite perfect yet.

Finally, keeping a hybrid system running smoothly requires constant attention. It's like having multiple high-maintenance machines. You need good monitoring systems and skilled people to make sure everything is working together and to catch any problems early on.

Despite these challenges, hybrid power is still a promising technology. As this technology improves and costs come down, these hurdles will become smaller. In the future, hybrid systems could be a major player in the clean energy game!

Case Study: Residential Hybrid System—The Smith Family, Colorado

Living off-grid in Colorado, the Smith family was tired of dealing with constant power outages and sky-high energy bills. They craved a reliable, eco-friendly solution—and that's where hybrid power swooped in to save the day.

Building Their Clean Energy Dream Team

Their system combined a 5 kW solar panel crew soaking up the sun, a 3 kW wind turbine catching the breeze, and a 10 kWh lithium-ion battery team for energy storage. The solar panels were strategically placed on the roof for maximum sun exposure, while the wind turbine found its perfect spot in an open area to catch those prevailing winds. A hybrid inverter kept everything in perfect balance, managing the flow of energy between all the players.

Powering Up With Sunshine and Wind

Since the big switch, the Smiths haven't missed a beat. Their hybrid system faithfully provides reliable power, even when the clouds roll in or the wind takes a vacation. This sunshine and wind dream team has slashed their energy costs and kicked their dependence on noisy diesel generators to the curb. The battery team also plays a key role, storing excess energy when the sun is shining bright or the wind is howling, and then doling it out when needed—making sure the Smiths never have to worry about running out of juice.

Going Green, One Watt at a Time

The environmental impact? A home run! By generating their own clean energy, the Smiths have shrunk their carbon footprint and become champions for a greener future. Their success story has even inspired their neighbors to explore the possibilities of hybrid power, turning their quiet corner of Colorado into a beacon of sustainable living.

Chapter 15:

Designing a Hybrid Power System

Hybrid System With Roof-Mounted Panels and Battery Backup Depiction

Hybrid systems are the ultimate team players in the renewable energy game, and this guide will show you how to build your own! We'll dive deep into everything you need to know to design a kick-ass hybrid system.

The real magic happens when we learn how to make these technologies work together seamlessly. You'll discover how to maximize the power you get from the sun and wind, ensuring a reliable and efficient energy flow. By the end, you'll be a master of resource harmony, ready to create a hybrid system that minimizes environmental impact while pumping out clean energy!

Assessing Combined Energy Potential

The first important step in creating a hybrid power system is to figure out how much energy solar and wind resources can produce together. This requires a thorough examination of data relevant to the site in order to ascertain whether combining these two renewable energy sources is possible and how well it is likely to work.

Data Collection and Analysis

To understand energy potential, it's important to collect accurate data. This means getting accurate weather information, like how much sunlight is hitting the earth and how fast the wind is blowing, from stable places like weather stations or monitors at the site. Wind data needs to show both short-term changes and long-term trends. Solar data needs to be taken for a year to cover all four seasons.

Once the data has been gathered, it needs to be studied to find out how much energy can be made. In the case of solar power, this means figuring out how much energy is expected to be produced based on average sun irradiance, panel efficiency, and system losses. For wind power, this means looking at patterns in wind speed and using models like the Weibull distribution to figure out how likely it is that different wind speeds will happen at the spot.

Complementarity of Resources

The synergy between wind and solar power is an important factor in the accumulated power output. When the sun isn't shining (during the day, for example, when clouds roll in), the wind should be picking up speed, and vice versa; this is how solar and wind resources should ideally work together. The hybrid system's energy production is more balanced and dependable because of their complementing nature.

Economic Feasibility

When evaluating the potential of combined energy sources, economic feasibility is also essential. To do this, you need to figure out the levelized cost of energy (LCOE) for both the wind and solar parts. After factoring in the initial investment, ongoing maintenance, and decommissioning expenses—the LCOE calculates the average cost per unit of power generated. To be financially viable, a hybrid system must outperform alternative energy sources in terms of LCOE.

Environmental and Regulatory Considerations

It is also important to look at the environmental effect and the regulatory framework. This means knowing how both solar and wind installations might affect the environment—such as how they might change the use of land, bother animals, and the appearance. Local rules about safety and permissions must be followed in order for a mixed system to be set up and run successfully. In order to do this, it may be necessary to get permits, do environmental impact studies, and follow zoning laws.

System Sizing and Configuration

To make sure that a hybrid power system works well and successfully for its intended purpose, it is essential to carefully consider its size and configuration.

Energy Demand Assessment

Finding out how much energy is needed is the first step in system sizing. This means looking at how much energy the house uses overall, including both peak and average use. Knowing how energy demand changes throughout the day and throughout the year can help you build a system that can provide enough power all year long. You can get this information from energy audits, electric bills, and smart meters that track real-time energy use.

Sizing Solar and Wind Components

After finding out how much energy is needed, the next step is to figure out how big the solar and wind parts of the combination system should be. In the case of solar power, this means figuring out how many solar panels are needed to meet the demand for energy, taking into account things like panel efficiency, solar radiation, and system losses. For wind power, the process includes picking the right models of wind turbines based on their power curves and how the wind blows at the spot.

When the system is being sized, it's crucial to make sure that the output from both the solar and wind panels is enough to meet the energy needs, even when production is low. Adding a safety gap, to account for sudden changes in energy production and demand, is a common way to do this.

Configuration and Optimization

Figuring out the best way to lay out and connect the solar and wind components is part of setting up the combination system. This includes where solar panels and wind turbines are placed so that they collect the most energy. The tilt and orientation of solar panels should be chosen to get the most sunshine. On the other hand, wind turbines should be placed to benefit from the direction of the wind and avoid obstacles that could cause turbulence.

Adding advanced control systems that handle the flow of energy between the solar panels, wind turbines, and the energy storage system is another way to make the configuration work better. These control systems use real-time data to change how the hybrid system works. This keeps the production and usage of energy in balance and stores any extra energy for later use.

Integration of Solar and Wind Components

Putting solar and wind power together in a hybrid system needs careful planning and execution to make sure it works smoothly and efficiently.

Electrical Integration

Putting together solar and wind power electrically means bringing the solar panels' and wind turbines' outputs to a single point. This is usually done with inverters, which change the DC output from the panels and the AC output from the turbines into a steady AC output. In these kinds of systems, hybrid inverters are often used because they can handle various energy sources and keep the power stable.

Energy Storage Integration

Energy storage is an important part of hybrid systems because it acts as a cushion, storing extra energy made during times of high production and releasing it during times of low production. Adding energy storage means picking the right battery technology and volume based on how much energy is needed and how it is made. The charging and discharging processes of the storage system should be set up to work best, ensuring longevity and efficiency.

Control and Monitoring Systems

Hybrid power systems need to be managed with the help of advanced control and tracking systems. Tracking the performance of the solar panels, wind turbines, and energy storage is done with the help of data from sensors and other tracking tools. They can set the system to automatically adjust its operation to balance energy production and usage, keep it from overheating, and make sure that energy storage is used effectively.

Control systems also give owners useful information for maintenance and repair, which helps them find and fix problems before they get worse. Monitoring features let you keep an eye on and handle the hybrid system in real time.

Hybrid System Integration Techniques

Integrating solar and wind power systems into one cohesive and efficient whole is the goal of hybrid system integration strategies.

Parallel and Series Configuration

In a parallel configuration, you connect the positive wires from the solar panels and wind turbines together, and you do the same with the negative wires. This setup boosts the amount of current (or amperage) the system can handle, but the voltage (the "push" behind the electricity) stays the same. The cool thing about this setup is that if one part stops working, the others can keep running without any problems. The arrangement maximizes dependability and versatility by allowing each component to operate independently.

In a series configuration, you connect the positive wire of one component to the negative wire of the next one, like forming a chain. This increases the voltage, which can be useful if your system needs a stronger electrical push. However, if one part of the chain breaks, the whole system stops working.

So, in simple terms: Parallel connections give you more current and are more reliable, while series connections give you more voltage but can be a bit riskier if something goes wrong.

The output of one part is used to power the input of another part in a series arrangement. The inverter of a wind turbine, for instance, might take the DC output of solar panels and transform it into AC for use by the load. Although this setup streamlines the control system, the extra conversion steps could lower overall efficiency.

Microgrids

Microgrids are small independent power lines that can work with the main grid or on their own. To add solar and wind power to a microgrid, you need to use complex control systems to oversee the flow of electricity between the different power sources and the energy storage system. Microgrids are more resilient and reliable than main grids because they can keep running even if the main grid goes down.

Safety Considerations for Hybrid Power Systems

It's important to make sure that combined solar wind power systems are safe. A complete safety plan includes the planning, setting up, using, and maintaining stages—looking at possible dangers and taking steps to stop them.

Design and Installation Safety

As part of the planning process, safety concerns include picking the right materials and parts that can handle the environment and electrical loads. To keep your electrical system safe during storms and avoid electrical hazards, you need to make sure you have proper grounding and lightning protection. Professionals who are certified and follow industry standards and state laws should do the installation. Making sure there is enough space and air flow keeps things from getting too hot and lowers the risk of fire.

Electrical Safety

In hybrid power sources, electrical safety is very important. Electrical problems are less likely to happen when there is good insulation, secure connections, and safety devices—like fuses and circuit breakers—are used. Ground fault protection (GFP) systems can find and separate ground faults, which keeps people from getting shocked by electricity and prevents electronics from breaking. Regular maintenance and checks of electrical parts help locate and repair problems before they get too bad.

Structural Safety

To be able to handle environmental stresses, like high winds and heavy snow loads, wind turbines and solar panels need to be mounted firmly. To keep mechanical problems from happening, mounting systems should have their structural stability checked on a regular basis. Accidents are less likely to happen if solar panels are firmly attached to wind turbines, and wind turbines are placed away from populated areas.

Fire Safety

When batteries are used in hybrid power sources, they can be a fire risk. Using fire-resistant materials, installing fire suppression systems, and keeping a safe distance around electrical equipment are all ways to prevent fires. Battery storage systems should have built-in thermal controls to keep them from getting too hot and starting a fire. Battery systems can detect early signs of thermal runaway when they are regularly checked and maintained.

Operational Safety

To run the hybrid system, safety rules should be in place while it's in use. Automatic control systems can keep an eye on how well a system is working and shut it down if something goes wrong, like too much electricity or heat. Owners should be taught how to handle situations, do regular checks, and strictly follow safety rules.

Maintenance and Monitoring

For long-term safety, it's important to do regular upkeep and checks. This includes checking all the parts on a regular basis, making sure the safety systems work, and replacing any old or broken parts right away. Advanced tracking systems give real-time information on how well a system is working, which lets problems be found early and maintenance to be done before issues occur.

Environmental and Human Safety

Safety concerns also include the health of people and the surroundings. Making sure that bird and bat deterrents are installed on wind blades helps protect wildlife in the area. Also, the noise level from the blades should be kept within acceptable limits so that it doesn't bother people who live nearby.

Chapter 16:

Maintaining Hybrid Systems

Hybrid System Depiction

Hybrid systems are superstars in the renewable energy world, but just like any machine, they need a little TLC to keep them running their best. This chapter will show you how to take care of your hybrid system, ensuring it delivers clean and reliable energy for years to come. By following a regular maintenance routine and dealing with any issues quickly, your hybrid system will keep pumping out clean energy for a long time. That way, you can enjoy all the benefits of renewable power without any worries!

Routine Maintenance for Hybrid Systems

Hybrid solar and wind power systems need to be serviced regularly to keep them safe, working well, and efficient. Because these systems have both solar panels and wind turbines, they need to be checked on a regular basis to make sure everything works right.

Solar Panel Maintenance

The solar panels need to be cleaned often. Depending on where you live, debris—such as dust, dirt, and bird droppings—can block sunshine and make the panels less effective. This means that the panels may need to be cleaned more than once a year. Make sure there are no cracks or scratches in the panels, and that all the electrical connections are tight and free of rust.

Battery Maintenance

The batteries in your hybrid system should be kept in good shape. Check the battery connections for corrosion on a regular basis. If you have lead-acid batteries, make sure the electrolyte levels are correct, and keep an eye on the charge and discharge cycles. Modern battery management systems can give you real-time information about the health of your batteries, which can help you find problems early.

Electrical System Checks

To keep everything safe and working well, the electrical system needs to be checked on a regular basis. This means checking the wires, inverters, and converters for damage, wear, and corrosion. Check to see if safety features, like fuses and circuit breakers, are working right. To keep electrical dangers from happening, check the grounding and bonding systems, as well.

Monitoring Systems

Modern hybrid systems often have high-tech tracking systems that show how much energy is being produced, how efficiently the system is working, and how healthy the parts are in real time. By looking over this data on a regular basis, you can spot trends and possible problems early on, which lets you do preventative maintenance. Keep the firmware and software updated on these tracking systems so that they have the newest features and security patches. Don't forget to do this.

Troubleshooting Hybrid System Issues

Hybrid systems need to be properly troubleshot in order to keep working well and being reliable. Finding problems and fixing them quickly can keep small issues from getting worse and leading to major fails.

Common Issues in Solar Components

One problem that solar panels can have is less power output because of things like shade, dirt, or damage. To fix this, you need to look at the panels and use tools like multimeters to check the electrical outputs. If a panel doesn't make nearly as much power as it should, it might need to be cleaned or replaced.

Wind Turbine Issues

Windmills can have problems with their mechanics and electricity. Problems with turbines often involve worn bearings, broken gearboxes, and damaged blades. Problems with electricity could include issues with the generator or the operating system. To fix these problems, you need to check them regularly, use diagnostic tools—like vibration monitors and thermal cameras—to find mechanical issues, and use the right testing equipment to check the electrical connections and outputs.

Battery and Electrical System Issues

Batteries can have problems like losing power, not being charged correctly, and getting too hot. As part of troubleshooting, the charge level, electrolyte levels, and termination connectors must be checked. Error codes from the system's tracking software can often help you figure out what's wrong with an electrical system, like when inverters or converters stop working. Making sure that all electrical connections are tight and testing parts with troubleshooting tools can help find and fix problems.

Control System Failures

Control systems are very important in mixed power setups for making sure that the system works properly and is integrated. If these systems break down, they might not work as well as they should, or could even shut down completely. When you troubleshoot, you have to make sure that all of the sensors and communication devices are working and that the control algorithms are being used properly. Problems might need to be fixed by updating and recalibrating software.

Upgrading and Expanding Hybrid Power Solutions

Improving mixed power systems makes them work better, produce more energy, and last longer. Compatibility and effectiveness depend on careful planning and execution.

Assessing the Need for Upgrades

First, look at the system's current performance, data on how much energy it produces, records of maintenance, and its general efficiency.

Upgrades might be needed to get rid of old parts, add new technologies, or increase the size of the system.

Adding Solar Panels or Wind Turbines

More solar panels or wind machines can be added to increase the amount of energy that is produced. Check the available room, the support for the structure, and the state of the environment. Make sure the electrical system can handle the extra load. Inverters, converters, and wires may need to be upgraded.

Enhancing Energy Storage

It is possible to make the system work better by adding or exchanging batteries with ones that hold more power, like lithium-ion or flow batteries. Integration with the current system is very important for smooth function.

Implementing Smart Grid Technologies

Add smart grid technologies to allow monitoring, control, and contact in real time. These can make the production and use of energy more efficient, keep the grid more stable, and make it easier to respond to changes in demand.

Improving System Efficiency With Advanced Inverters

These days, hybrid transformers can handle more than one energy source and store energy better. Features like maximum power point tracking (MPPT) help keep the power grid stable and get the most energy from solar cells.

Software and Firmware Updates

Control systems, inverters, and tracking devices need to have their software and firmware updated on a regular basis to keep them running, add new features, and make them safer.

Regulatory and Environmental Considerations

Make sure that local rules and environmental standards are followed, get the permits you need, and do environmental effect assessments. Getting involved with local leaders early on can make the execution of a project go more smoothly.

Economic Analysis and Financing

Do a full economic analysis to figure out how much the new parts, installation, and possible downtime will cost. Think about the projected rise in energy production and the money you will get back. Find rewards—such as rebates and tax credits—to help pay for the start-up costs.

Project Management and Execution

Planning, organizing, and coordinating in great detail are all parts of good project management. Hire contractors and suppliers with a lot of knowledge to make sure you get good work, and stay on budget and on time.

Post-Upgrade Testing and Monitoring

After upgrades, make sure the system works correctly by trying it thoroughly and keeping an eye on it. Test for performance, make sure electrical links are correct, and set up control systems. Monitoring the system all the time helps find problems quickly and makes sure it meets its energy production goals.

Conclusion

We've reached the end of our off-grid lessons, and it's become clear that this lifestyle is about so much more than just wires and switches. It's a path to independence, self-reliance, and a closer bond with the natural world.

This book has been your guide on this journey, exploring everything for generating your own electricity. But now, let's take a step back and consider the bigger picture. How will living off the grid affect you and the environment around you?

Embracing Independence

Living Off-Grid offers a powerful kind of independence. You become the master of your own domain by generating your own electricity. This shift isn't just about flipping switches; it's about building self-reliance and resilience.

Power outages? No sweat. You'll have the skills and resources to meet your needs, fostering a deep sense of empowerment. It's a chance to take control and become less reliant on external systems. This self-sufficiency is one of the most compelling aspects of living off the grid.

Environmental Impact

Living off the grid means you're doing your part for the environment. By switching to renewable energy sources, like solar and wind, you're cutting down on your carbon footprint in a big way. This helps fight climate change and supports a more sustainable lifestyle. Living off-grid pushes you to conserve and use resources wisely, promoting a mindset that really values the planet and all it offers.

Financial Benefits

Sure, the upfront cost of going off-grid can be pretty hefty, but the long-term financial perks are worth it. Over time, you'll notice a big drop in your utility bills. Savings on electricity, water, and even food can really stack up—making living off-grid a smart financial move. Plus, the skills you pick up—like gardening, building, and maintenance—can lead to even more savings and might even open up new income opportunities (which are topics detailed in our upcoming books).

Building Community

Living off-grid might seem like a lonely path, but it often leads to forming tight-knit, supportive communities. Many off-gridders create networks to share knowledge, resources, and support. Whether it's through local groups or online forums, these communities are great for troubleshooting, sharing wins, and making friends with people who get it. This sense of community is super rewarding and really helps you stick with the off-grid lifestyle.

Personal Growth

Going off-grid is also a journey of personal growth. You'll need to pick up new skills, tackle challenges, and keep tweaking and improving your systems. This constant learning can be really fulfilling. You'll start appreciating things you used to take for granted and gain a deeper understanding of how stuff works, from electrical systems to sustainable farming practices.

Challenges and Rewards

Living off-grid definitely comes with its challenges. It takes careful planning, hard work, and a knack for troubleshooting and adapting. But the rewards make it all worth it. There's nothing quite like the satisfaction of generating your own power, growing your own food, and living in harmony with nature. Every challenge you conquer proves your resourcefulness and determination, making the whole experience incredibly rewarding.

Looking Ahead

As you start or keep going on your off-grid journey, remember that it's all about continuous improvement and adapting. Technologies will keep evolving, and so will your needs and goals. Stay updated on the latest in renewable energy, sustainable practices, and off-grid living techniques. Embrace these changes and stay open to learning and growing along the way.

Final Thoughts

Living off the grid isn't just a lifestyle choice; it's a commitment to a way of life that values sustainability, independence, and a deep connection to nature. It's about making intentional choices that reflect your values and priorities. Whether you're driven by environmental concerns, a desire for independence, or just looking for a simpler, more fulfilling life—*Living Off-Grid* can help you achieve those goals.

As you wrap up this book, take with you the knowledge, inspiration, and practical advice it offers. Let it be your guide on the journey to a more self-sufficient and sustainable future. The possibilities are endless, and the rewards are profound. Welcome to the world of *Living Off-Grid*, where each step brings you closer to a life that's not only sustainable but also deeply rewarding.

We hope this book charged you up with inspiration and knowledge. If you enjoyed it, we'd love for you to spark a review. Your feedback helps us produce such intriguing books; it keeps us going!

Thank you for joining us on this journey. Here’s to a brighter, greener, and more independent future!

Glossary

- **Alternative Energy:** Energy created so that it does not use up natural resources or cause environmental damage, particularly by using renewable sources—such as wind, solar, and hydropower.
- **Amperage:** The force of an electrical current.
- **Autonomy:** The capacity to make an independent decision and function independently without relying on external systems or support.
- **Battery Storage:** Systems used to store energy produced by renewable sources for use at a later time, ensuring a consistent power supply.
- **Biodiversity:** The variety of plant and animal life in a particular habitat, crucial for preserving the balance and health of ecosystems.
- **Carbon Footprint:** The sum total of greenhouse gases, primarily carbon dioxide, emitted directly and indirectly by human activities.
- **Climate Change:** Long-term alterations in temperature, precipitation, wind patterns, and other aspects of the Earth's climate system, largely due to human activities such as burning fossil fuels.

- **DC-AC Conversion:** The process of converting direct current (DC) electricity, generated by solar panels, into alternating current (AC) electricity, used by most household appliances.

- **Energy Efficiency:** The goal of using less energy to deliver the same service, reducing energy waste and saving costs.

- **Flywheels:** Rotating disc that stores energy for later use (kinetic energy storage).

- **Gin Pole:** A mast bolted to the tower with a pulley or block tackle to lift loads and is used to lift the next tower section into position.

- **Grid-Tied Systems:** Renewable energy systems that are connected to the public electricity grid, allowing for excess energy to be returned to the grid.

- **Horizontal Axis Wind Turbine (HAWT):** A type of wind turbine where the main rotor shaft is set horizontally, parallel to the ground—and the blades rotate perpendicular to the wind flow.

- **Hybrid Inverter:** A device that can maneuver power from the grid, solar panels, and batteries—ensuring a consistent power supply and optimizing energy usage.

- **Inverter:** A device that converts DC electricity produced by renewable energy sources into AC electricity for use in homes and businesses.

- **Maximum Power Point Tracking (MPPT):** A tech device that tracks in real time the power output of solar panels and turbines in order to increase power in battery storage.

- **Micro-Hydro Systems:** Small-scale water-powered systems used to generate electricity, often used in off-grid living.

- **Off-Grid:** A lifestyle characterized by not being connected to the public electricity grid, relying instead on self-sufficient energy and water sources.

- **Photovoltaic (PV) Cells:** Semiconductor devices help derive electricity from sunlight, using solar panels.

- **Pumped Hydro Storage:** A type of hydroelectric power generation used for storing energy by pumping water from a lower height reservoir to a higher height. During high electricity demand, the stored water is released back down to the lower reservoir, transversing turbines to generate electricity.

- **Renewable Energy:** Energy from natural sources—such as sunlight, water, and wind—that are refilled repeatedly.

- **Self-Sufficiency:** Being able to satisfy all of one's needs without outside help, particularly regarding food, energy, and water.

- **Smart Grids:** An electricity system that uses digital tech to keep track of and manage how electricity moves from where it's made to where it's needed, making sure everyone gets the power they need.

- **Solar Panels:** Devices composed of photovoltaic cells that change sunlight into electricity.

- **Sustainable Living:** A way of life focused on cutting down on how much of the planet's resources and personal resources we use.

- **Thermal Management:** The regulation of temperature in renewable energy systems, particularly in batteries, to ensure efficient operation and longevity.

- **Upwind Fetch:** The distance over which the wind blows across a body of water, without obstruction, which influences the formation of waves and the amount of energy a wind turbine can harness.

- **Vertical Axis Wind Turbine (VAWT):** A type of wind turbine where the main rotor shaft is vertical, perpendicular to the ground.

- **Voltage:** The force from an electrical circuit's power source that drives the current to a device.

- **Wind Turbines:** Devices that transform kinetic energy from the wind into mechanical power or electricity.

References

Abeni, M. (2024, July 15). *How efficient are solar panels?* Sunsave. https://www.sunsave.energy/solar-panels-advice/how-solar-works/efficiency

AleaSoft. (2023, September 29). *The drop in the LCOE of renewable energies over the past decade drives the energy transition.* https://aleasoft.com/drop-lcoe-renewable-energies-past-decade-drives-energy-transition/

The Betz limit for wind turbine power. (n.d.). Alternative Energy Tutorials. https://www.alternative-energy-tutorials.com/wind-energy/betz-limit.html

Bhatia, A. (n.d.). *Wind energy design and fundamentals* [Mooc]. CEDengineering.com. https://www.cedengineering.com/userfiles/R06-004%20-%20Wind%20Energy%20Design%20and%20Fundamentals%20-%20US.pdf

Bifacial solar panels: What are they and are they worth it? (2024, April 14). *Alternergy.* https://www.alternergy.co.uk/blog/post/bifacial-solar-panels-what-are-they

Chodos, A. (2005, January 1). *Einstein and The Photoelectric Effect.* American Physical Society (APS). https://www.aps.org/apsnews/2005/01/einstein-photoelectric-effect

Chodos, A. (2009, April 1). *April 25, 1954: Bell Labs demonstrates the first practical silicon solar cell.* American Physical Society (APS). https://www.aps.org/apsnews/2009/04/bell-labs-silicon-solar-cell

Christian M. (2024, August 7). The rise of flow batteries: A new era. *AquaSwitch.* https://www.aquaswitch.co.uk/blog/flow-batteries/

Crystal. (2018, April 1). *What is MPPT (maximum power point tracking).* VEICHI. https://www.veichi.com/knowledge/what-is-mppt.html

Dodds, P. E., & Garvey, S. D. (2016). The role of energy storage in low-carbon energy systems. In T. M. Letcher (Ed.), *Storing Energy* (pp. 3-22). Elsevier. https://doi.org/10.1016/B978-0-12-803440-8.00001-4

The Editors of Encyclopædia Britannica. (2019). Sundial. In *Encyclopædia Britannica.* Retrieved July 23, 2024 https://www.britannica.com/technology/sundial

The Editors of Encyclopædia Britannica. (2020). Horizontal-axis wind turbine. In *Encyclopædia Britannica.* Retrieved July 23, 2024 https://www.britannica.com/technology/horizontal-axis-wind-turbine

Edmond Becquerel: The man behind solar panels. (2019 October 24). Solenergy Systems. https://solenergy.com.ph/solar-panel-philippines-edmond-becquerel/

Energy Education Team. (n.d.). *Betz limit.* Energy Education. https://energyeducation.ca/encyclopedia/Betz_limit

Fitton, K. (2024, March 7). *Storing renewable electricity on the grid of the future (updated for 2024).* Medium.

https://medium.com/@kitfitton/storing-renewable-electricity-on-the-grid-of-the-future-updated-for-2024-ff344531eaeb

Foxtron. (2021, December 3). *Depth of discharge: What it is and why it's important.* Foxtron Power Solutions. https://www.foxtronpowersolutions.com/depth-of-discharge/

Geotherm_Admin. (2017, April 29). *Polycrystalline solar cells vs monocrystalline solar cells.* Geotherm. https://geothermhvac.com/mono-vs-poly-better/

Hansen, J. T., Mahak, M., & Tzanakis, I. (2021). Numerical modelling and optimization of vertical axis wind turbine pairs: A scale up approach. *Renewable Energy*, *171*, 1371–1381. https://doi.org/10.1016/j.renene.2021.03.001

Horizontal axis wind turbines: Advantages and disadvantages. (2020, March 31). *LuvSide.* https://www.luvside.de/en/hawt-advantage-disadvantage/

How solar panel efficiency and cost changed over time. (2024, July 10). Renogy. https://www.renogy.com/about-us/

Howard, B. C. (2017, January 13). Iran's centuries-old windmills may soon stop turning. *National Geographic.* https://www.nationalgeographic.com/science/article/nashtifan-iran-windmills

Lane, C. (2024, June). How to wire solar panels in series vs. parallel. *SolarReviews.* https://www.solarreviews.com/blog/do-you-wire-solar-panels-series-or-parallel

Magdalin, A. E., Nixon, P. D., Jayaseelan, E., Sivakumar, M., Devi, S. K. N., Subathra, M. S. P., Kumar, N. M., & Ananthi, N. (2023). Development of lead-free perovskite solar cells: Opportunities, challenges, and future technologies. *Results in Engineering, 20*, 101438. https://doi.org/10.1016/j.rineng.2023.101438

Perlin, J. (n.d.). *The story of Vanguard.* UCSB Experimental Cosmology Group. https://www.deepspace.ucsb.edu/outreach/the-space-race/the-story-of-vanguard

Ranjbar, M., Nasrazadani, S., Kia, H., & Gharali, K. (2019). Energy Equipment and Systems Reaching the betz limit experimentally and numerically. *Energy Equip. Sys*, *7*(3), 271–278. https://www.energyequipsys.com/article_36563_614f7306df0055ced9d46515474eba30.pdf

Reiss, A.J., Figurskey, D., Bassett, C., & Cuff, T. (2021). Chapter 5 - Propelling the new blue economy with safer marine transportation and infrastructure. In L. Hotaling, & R. W. Spinrad (Eds.), *Preparing a workforce for the new blue economy* (pp. 63-84). Elsevier. https://doi.org/10.1016/B978-0-12-821431-2.00003-2

Shahan, Z. (2014, November 21). *History of wind turbines.* Renewable Energy World. https://www.renewableenergyworld.com/storage/grid-scale/history-of-wind-turbines/

天空 (Sky). (2024, February 7). Battery longevity secrets: How long do lithium batteries last? *Lithium Battery Tech.* https://www.lithiumbatterytech.com/how-long-do-lithium-batteries-last/

Solar Optimum. (2023, January 11). What are the advantages of a monocrystalline solar panel? *Solar Optimum.* https://solaroptimum.com/blog/2023/01/11/what-are-the-advantages-of-a-monocrystalline-solar-panel/

Thin film vs. crystalline solar panels: Which one is better for your home? (2023, August 8) SolarSME. https://solarsme.com/thin-film-vs-crystalline-solar-panels/

US Department of Energy. (2019). *History of U.S. wind energy*. Office of Energy Efficiency and Renewable Energy (EERE). https://www.energy.gov/eere/wind/history-us-wind-energy

Vector Renewables. (2023, February 1). Do you know what the 'wake effect' is in a wind farm? *Vector Renewables.* https://www.vectorenewables.com/en/blog/do-you-know-what-the-wake-effect-is-in-a-wind-farm

Images Reference

Angel, R. G. (2019, May 31). *Two men carrying a metal sheet* [Image]. Unsplash. https://unsplash.com/photos/two-men-carrying-a-metal-sheet-MagdWoazARo

Arantes, L. (2022, October 9). *A yellow windmill against the blue sky* [Image]. Pexels. https://www.pexels.com/photo/a-yellow-windmill-against-the-blue-sky-14700028/

Bizar, J. V. (2023, May 30). *Solar panels on house roof* [Image]. Pexels. https://www.pexels.com/photo/solar-panels-on-house-roof-17762230/

Derndorff, M. (2024, April 23). *Clouds over a windmill standing in a field* [Image]. Pexels. https://www.pexels.com/photo/clouds-over-a-windmill-standing-in-a-field-22491137/

Gremlin. (2023, February 10). *Modern house with solar photovoltaic roof panels and wind turbine* [Image]. IStock. https://www.istockphoto.com/photo/modern-house-with-solar-photovoltaic-roof-panels-and-wind-turbine-gm1464571552-497235876?utm_source=pixabay&utm_medium=affiliate&utm_campaign=SRP_image_sponsored&utm_content=https%3A

%2F%2Fpixabay.com%2Fimages%2Fsearch%2Fhome%2520wind%2520turbines%2F&utm_term=home+wind+turbines

Kuznetsova, K. (2022, December 26). *A man standing next to a solar panel* [Image]. Unsplash. https://unsplash.com/photos/a-man-standing-next-to-a-solar-panel-InYNrJD_zfY

Mead, B. (2021, June 22). *Man in white dress shirt and blue denim jeans sitting on white and black solar panel* [Image]. Unsplash. https://unsplash.com/photos/man-in-white-dress-shirt-and-blue-denim-jeans-sitting-on-white-and-black-solar-panel-wmaP3Tl80ww

Nikitina, V. (2023, March 2). *A house with a solar panel on the roof next to a wind turbine* [Image]. Unsplash. https://unsplash.com/photos/a-house-with-a-solar-panel-on-the-roof-next-to-a-wind-turbine-0Ih8YLg2bKM

Oroni, P. (2024, May 30). *Three houses connected to a battery with windmills in the background* [Image]. Unsplash. https://unsplash.com/photos/three-houses-connected-to-a-battery-with-windmills-in-the-background-kIlYKwJZ-bM

Pixabay. (2017, April 10). *White concrete house surrounded by trees* [Image]. Pexels. https://www.pexels.com/photo/white-concrete-house-surrounded-by-trees-414911/]

Sloot, J. (2019, August). *Black and white wind mill in green field under yellow skies.* Pexels [Image]. https://www.pexels.com/photo/black-and-white-wind-mill-in-green-field-under-yellow-skies-2786784/

Unsplash+ In collaboration with Getty Images . (2023, April 22). *Installing solar panels, close-up on a working tools. wires and man in protective clothing standing on a rooftop with photovoltaic power* [Image]. Unsplash. https://unsplash.com/photos/installing-solar-panels-close-up-on-a-working-tools-wires-and-man-in-protective-clothing-standing-on-a-rooftop-with-photovoltaic-power-station-DLemF19LnzM

Unsplash+ In collaboration with Resource Database. (2024, May 30). *A group of green cars parked next to a wind turbine* [Image]. Unsplash. https://unsplash.com/photos/a-group-of-green-cars-parked-next-to-a-wind-turbine-Hc6hQa8aAOc

Unsplash+ In collaboration with Getty Images. (2022, August 31). *Technologist in a white coat makes the necessary entries in the tablet is at the factory* [Image]. Unsplash. https://unsplash.com/photos/technologist-in-a-white-coat-makes-the-necessary-entries-in-the-tablet-is-at-the-factory-the-man-selects-the-desired-parameters-on-the-digital-touch-screen-on-the-device-in-the-shop-worker-operating-pasteurizer-using-the-control-panel-KhikHsrQGT4

User, P. (n.d.). *A windmill is sitting on a dock next to a body of water* [Image]. Pexels. https://www.pexels.com/photo/a-windmill-is-sitting-on-a-dock-next-to-a-body-of-water-25361754/

Other images provided by authors.

Author Bio

JD and Twila Weiss are the hardworking co-owners of "Rebel Ranch Publishing" and the "Rebel Ranch." These talented individuals operate a Belted Galloway cattle ranch and a Honeybee Apiary. Together, JD and Twila have achieved recognition and book sales not only in the United States but also on an international scale.

Their diverse collection of books can be easily obtained from online platforms, local physical bookstores, and libraries. With years of writing under their belt, they bring a unique perspective to the non-fiction, fiction, and children's book genres. Moreover, their skillset extends beyond traditional writing, encompassing the creation of official documents and innovative papers that set them apart in their field.

Writing Expertise

One cannot help but be impressed by the depth of experience exhibited by JD and Twila Weiss. Their journey as published authors has given them valuable insights into the intricacies of the written word. From shaping compelling narratives in fiction to elucidating complex concepts in non-fiction, the dynamic duo does it all with finesse. Their ability to tackle various writing styles and genres reflects their adaptability as writers.

Ongoing Projects

JD and Twila are fully immersed in the creative process, diligently working on multiple book manuscripts. This highlights their dedication to their craft and their unyielding commitment to producing high-quality content consistently. It is this passion for storytelling that drives them forward, pushing the boundaries of their creativity to deliver captivating narratives that resonate with readers across the globe.

Made in United States
North Haven, CT
04 November 2026

21769526R00118